D1665942

Petra Reinke

Inorganic Nanostructures

Related Titles

Schulte, J. (ed.)

Nanotechnology

Global Strategies, Industry Trends and Applications

2005

ISBN: 978-0-470-85400-6

Reich, S., Thomsen, C., Maultzsch, J.

Carbon Nanotubes

Basic Concepts and Physical Properties

2004

ISBN: 978-3-527-40386-8

Kelsall, R., Hamley, I. W., Geoghegan, M. (eds.)

Nanoscale Science and Technology

2005

ISBN: 978-0-470-85086-2

Waser, R. (ed.)

Nanoelectronics and Information Technology

Advanced Electronic Materials and Novel Devices

2003

ISBN: 978-3-527-40363-9

Petra Reinke

Inorganic Nanostructures

Properties and Characterization

WILEY-VCH Verlag GmbH & Co. KGaA

The Author

Prof. Petra Reinke
University of Virginia
Dept. of Mat. Science and Eng.
395, McCormick Road
Charlottesville, VA 22904
USA

Library of Congress Card No.: applied for

British Library Cataloguing-in-Publication Data
A catalogue record for this book is available from the British Library.

Bibliographic information published by the Deutsche Nationalbibliothek
The Deutsche Nationalbibliothek lists this publication in the Deutsche Nationalbibliografie; detailed bibliographic data are available on the Internet at <http://dnb.d-nb.de>.

Typesetting MPS Limited, a Macmillan Company, Chennai, India
Printing and Binding Strauss GmbH, Mörlenbach, Germany
Cover Design Adam-Design, Weinheim, Germany

Printed in Germany
Printed on acid-free paper

Print ISBN: 978-3-527-40925-9
ePDF ISBN: 978-3-527-64593-0
oBook ISBN: 978-3-527-64590-9
ePub ISBN: 978-3-527-64592-3
Mobi ISBN: 978-3-527-64591-6

Contents

Inorganic Nanostructures: Properties and Characterization, First Edition. Reinke, P.

Published 2012 by WILEY-VCH Verlag GmbH & Co. KGaA

Preface

Writing this book was an adventure - sometimes more like a rollercoaster ride, often peppered with surprising discoveries, and infused with many fascinating ideas and technological developments. Nanoscience has in the last decade evolved into one of the fastest paced areas in science and has had an impact on nearly every discipline including materials science, medicine, physics, chemistry, and biology. It is a truly interdisciplinary endeavor, which is not only evident in the pertinent literature but also increasingly visible in the classroom. The number of publications in many areas of nanoscience and -technology is increasing rapidly, one recent example is the discovery of graphene: in the year 2000 the number of publications on graphene was small, just a few people growing "dirt" on metal surfaces, but with the recognition of its extraordinary electronic properties, interest in this unique material exploded. By now it has become nearly impossible to even keep up with the literature. But graphene is also an old material: the first attempts to understand bonding in graphite used graphene as a model system without realizing that it is indeed possible to make these single layers. Understanding of properties combined with the newly found ability to synthesize the material created the "perfect storm" a few years ago. The present book ventures to illustrate this correlation: how can we make a material and how can we understand and then control its properties? What can we learn by understanding synthesis, how can we achieve the superb control over structure, geometry, composition which is needed to make fully functional nanostructures? I hope it will guide students and researchers alike on a journey into a wonderful and ever expanding world of nano-materials.

A large project such as writing a book or conducting research cannot be done alone, and this is the place to thank everybody who has supported me throughout my career. As you might have guessed, the list is long, and today I limit myself to mention just a few, otherwise quite a few more pages would have to be added to this book. First and foremost my thanks go to my students, past and present, and those who share currently in the exciting exploration of surfaces and nanomaterials (and proofread several chapters) - you will find some of their work in the pages of this book and on

Inorganic Nanostructures: Properties and Characterization, First Edition. Reinke, P.

Published 2012 by WILEY-VCH Verlag GmbH & Co. KGaA

the cover page. Their comments on my book manuscript and our discussions in the lab, office, classroom and during "boot-camp sessions" were (and are) instrumental in shaping my thoughts and ideas. Thanks to all my colleagues with whom I share discussions and thoughts on science and teaching and life, and those who proofread parts of this book - Archie Holmes, and Renee Diehl - your comments were critical to the development of my thoughts. I was fortunate to have in Bill Johnson a department chair who gave me the opportunity to develop and teach several courses on Nanoscience, which in the end culminated in the writing of this book.

Acknowledgements for Cover Art

All images displayed on the cover are STM (scanning tunneling microscopy) images taken by students in my research group. They illustrate their skill and dedication, and showcase some of the projects we have pursued in recent years.

Center image: Ge quantum dot (hut) grown by Stranski-Krastanov growth of Ge on the Si(100) surface. The QD was fabricated and recorded by Christopher A. Nolph in the framework of an NSF award by the Division of Materials Research (Electronic and Photonic Materials) award number DMR-0907234.

Image on left hand side (green): Surface of a fullerene layer - the spheres are individual fullerene molecules, which rotate at room temperature and therefore the individual atoms cannot be distinguished. The fullerene layer was deposited and imaged by Harmonie Sahalov. Her project was supported by NSF award number DMR-105808 (Division of Materials Research, Ceramics).

Center image - back panel (blue): This image shows the structures formed by Vanadium metal if it is deposited on a graphite surface at room temperature. The deposition and imaging were done by Wenjing Yin, and her project was supported by the Defense Microelectronics Agency under contract DMEA2-H94003-08-2-0803.

Image on right hand side (yellow): The surface depicted in this image is a Si(100)(2x1) reconstructed surface with mono-atomic Manganese wires, which run perpendicular to the Si-dimer rows. The work was done by Kiril R. Simov in the framework of NSF awards by the Division of Chemistry (Electrochemistry and Surface Chemistry) CHE-0828318, and Division of Materials Research (Electronic and Photonic Materials) DMR-0907234. This image is published in Simov, K. R., Nolph, C.A., and Reinke, P. (2012) Guided Self-assembly of Mn-Wires on the Si(100)2×1) Surface in *J. Phys. Chem. C* **116**, 1670. Reprinted with permission, copyright 2011 American Chemical Society.

1
Dimensions and Surfaces – an Introduction

This first chapter can be seen as a warm-up: it will prepare our mental muscles to think about nanomaterials, and why they can be considered as a class of materials in their own right. We will introduce the concept of confinement and dimensionality and derive the density of states (DOS) for low-dimensional structures. After a discussion of electronic properties we will move on to a quite different area of research, and discuss fundamental processes at surfaces, which are rarely included in materials science or physics core classes, but are important for the understanding of many aspects of nanomaterial synthesis.

1.1
Size, Dimensionality, and Confinement

The nanosize regime is defined by the transition between the bulk and atom, and is characterized by a rapid change in material properties with size. Each set of properties (mechanical properties, geometric and electronic structure, magnetic and optical properties, and reactivity) is defined by characteristic length scales. If the size of the system approaches a characteristic length scale, the property in question will be modified dramatically as a function of size. The intimate link between size and material properties is one of the most intriguing aspects of nanoscience, and is at the core of the discipline. The control of size is therefore often the most important, and difficult, challenge in the synthesis of nanostructures.

The decrease in size of a nanostructure is accompanied by a rapid change in the volume-to-surface ratio of atoms: a cube with a side length of 1 mm contains about $2.5 \cdot 10^{19}$ atoms, and the percentage of surface atoms is only $2 \cdot 10^{-6}$; for a cube side length of 1 μm the percentage of surface atoms increases to $2 \cdot 10^{-3}$; and for 1 nm side length, only one true volume atom remains, which is surrounded on all sides by other atoms. This shift from a volume–atom dominated structure, where the majority of atoms has fully

Inorganic Nanostructures: Properties and Characterization, First Edition. Reinke, P.

Published 2012 by WILEY-VCH Verlag GmbH & Co. KGaA

saturated bonds, to a surface–atom dominated structure has rather dramatic consequences.

One of the best-known examples, which illustrates the impact of the change in the ratio of surface-to-volume atoms, is the observation of the reactivity of nanosize catalyst particles [1–3]. Catalysts are industrial materials, which are produced in very high volumes and used in nearly every chemical process. The role of a catalyst in a chemical reaction is to lower the activation energies in one or several of the reaction steps, and it can therefore increase yield, reaction speed, and selectivity. Most catalysts contain a relatively high percentage of expensive noble metals, and increasing catalyst efficiency through reduction of its size can thus greatly diminish costs, and at the same time very often boosts efficiency. The reactivity increase with decreasing particle size can be attributed to several size dependent factors: a proportional increase in the number of reactive surface atoms and sites, changes in the electronic structure, and differences in the geometric structure and curvature of the surface, which presents a larger concentration of highly active edge and kink sites. The underlying mechanism of a catalytic reaction is often complex, and cannot be attributed to a single factor such as larger surface area or modulation of the electronic structure. The study of catalysts and catalytic reactions is a highly active field of research, and depends on the improved comprehension of nanoparticle synthesis and properties.

An important step in classifying the functionality of nanostructures is to understand the relation between dimensionality and confinement. Dimensionality is mathematically defined by the minimum number of coordinates required to define each point within a unit; this is equivalent to vectors which define a set of n unit vectors required to reach each point within an n-dimensional space. When looking at nanostructures, the definition of dimensionality becomes more ambiguous: for example, a semiconductor nanowire can have a diameter of a few to several ten nanometers, with a length up to several micrometers. It is a structure with a very high aspect ratio, but in order to describe the position of each atom or unit cell within the wire, a three-dimensional (3D) coordinate system is required with one axis along the wire and the other two unit vectors to describe the position within the horizontal plane. This coordinate system bears no relation to the crystal structure and only serves to illustrate the mathematical dimension of the nanowire. A one-dimensional (1D) nanowire is therefore strictly speaking only present if its thickness is only a single atom. Examples for this kind of 1D system are given in Chapter 3.

The most important aspect for our discussion of dimensionality is the modulation of the electronic structure as a function of the extension of a nanostructure in the three dimensions of space. It is possible to define potential barriers in a single direction in space, thus confining electrons in one direction, but leaving them unperturbed in the other two directions. This corresponds now to a two-dimensional (2D) nanostructure, a so-called quantum well. Going back to our example of the nanowire: the electronic system of the nanowire (if it is sufficiently small) is confined in the two directions perpendicular to its long axis, but not along the long axis itself, and it is therefore a 1D structure. The dimensionality of a nanostructure is defined through the geometry of the confinement potential.

Dimensionality for nanostructures is therefore often defined in a physically meaningful manner by considering the directions of electron confinement. Confinement for electrons is introduced in quantum mechanics by using the particle in a box: the electron wave is confined within the well, which is defined by infinitely high potential energy barriers. The equivalent treatment can be used for holes. The width of the box then controls the energy spacing between the allowed states, which are obtained from solutions of the Schrödinger equation.

The allowed n-th energy level E_n is given for a 1D well (a one dimensional box, with only one directional axis) by:

$$E_n = \frac{\hbar^2 n^2 \pi^2}{L^2 2m^*} = \frac{h^2 n^2}{8m^* L^2} \tag{1.1}$$

where L is the width of the box, m^* is the effective mass of the electron, and n is an integer 1, 2, 3, ... This equation emerges directly from the solution of the Schrödinger equation for the free electrons in a 1D box with a width of L, which is described in quantum mechanics textbooks. The energy increases with the inverse of the square of the box size, and n is the corresponding quantum number. This relation is quite general, and specific factors and exponents are modified by the shape of the confinement potential. The highest filled level at a temperature of 0 K corresponds to the Fermi energy (E_F) and the quantum well is in its ground state when all levels up to E_F are filled. The band gap is the energy difference between the ground state and the first excited state when one electron is excited to the first empty state above E_F. If we build a very large quantum well whose dimension approaches that of a macroscopic solid, the energy difference between the ground state and excited state will become infinitesimally small compared with the thermal energy, and the band gap created by quantum confinement for small L disappears; we now have a quasi-continuum of states. The height of the well barrier for a solid is given by the work function of the limiting surfaces.

The band gap in a macroscopic solid forms due to the periodicity of the lattice, which imposes boundary conditions on the electron waves and leads for certain energies to standing waves within the lattice. The standing waves whose wavelengths correspond to multiples of interatomic distances in a given lattice direction define the band gap within the band structure ($E(\vec{k})$) of the material (we are neglecting any structure factors on this discussion). The ion cores define the position of the nodes, and extrema of a standing wave. If the wave vector $\vec{k}$ satisfies the Laue diffraction condition within the reciprocal lattice, we will observe opening of a band gap at this specific value of $\vec{k}$, which is the Brillouin zone boundary. The energy gap opens due to the energetic difference between a wave where the nodes are positioned at the ion cores, and a wave of the same wavelength (wave vector) but where the nodes are positioned in between the ion cores. This argument follows the so-called Ziman model and is described in detail in many textbooks on solid state physics. A semicondcutor or insulator results if the Fermi energy E_F is positioned within this bandgap, in all other cases when E_F is positioned in the continuum of states, we will have a metal.

If we now start with a metal, where E_F lies within the continuum of states, and reduce the size of the system to the nanoscale, which is equivalent to the reduction in the size of the box or quantum well L, the energy difference between states will increase and we can open a band gap for sufficiently small dimensions. The macroscopic metal can become a nanoscopic insulator. For a semiconductor, where a fundamental gap is already present, the magnitude of the gap will increase as confinement drives the increase in separation of the energy levels. The measurement of the magnitude of the band gap is therefore a sensitive measure for quantum confinement and is used in all chapters to illustrate and ascertain the presence of confinement.

In the case of a 2D quantum well, if we define the confinement potential along the z axis, the band structure in x, and y will not be perturbed by confinement. Figure 1.1 shows the energy levels in the quantum well in the direction of confinement, z, and illustrates the sub-band formation. The confinement as described in Equation 1.1 only affects the energy levels in the z direction, while the continuum of levels in the x, and y directions is preserved and for the free electron case the dispersion relation is described by a parabola. Each discrete energy level (set of quantum numbers) in the directions of confinement is associated with the energy levels in the other directions; this leads to the formation of sub-bands. This train of thought can be transferred directly to quantum wires, where confinement is in two directions, and quantum dots, where confinement is in all three directions of space and we have a zero-dimensional (0D) electronic structure. Confinement and formation of sub-bands can substantially change the overall band structure, which is discussed for Si nanowires in the context of the blue shift of emission for porous silicon and a transition from an indirect to direct band gap material for small wire diameters [4–6] (see Chapter 4). The size of the gap is a signature of the impact of confinement, and its increase for decreasing size of a nanostructure is illustrated for several types of materials throughout this book. Silicon nanowires and graphene nanoribbons are examples where the increase in the gap was observed experimentally as a function of size.

The confinement potential will only affect the electronic structure if the dimensions of the potential well are in the range or smaller than the de Broglie wavelength, λ, which is given by:

$$\lambda = \frac{h}{(2m^* E)^{1/2}} \tag{1.2}$$

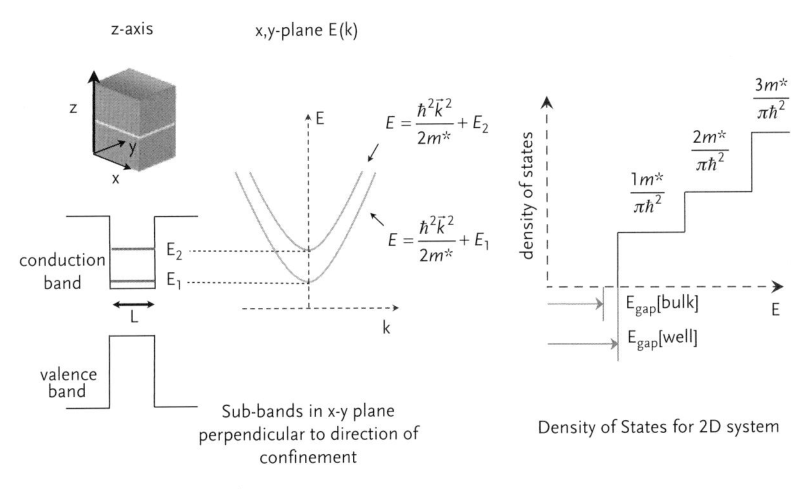

Figure 1.1 The discussion of confinement for a 2D quantum well with one direction of confinement (z axis) is summarized. The QW is sketched on the left hand side: a material with a smaller band-gap (light stripe) is embedded in a material with a larger band-gap (grey). The left-hand figure shows the energy levels within the quantum well, for clarity of illustration only two energy levels are included in the conduction band. The figure in the center shows the two parabolic sub-bands and the total energy of the electrons in these sub-bands, which is composed of the parabolic and the quantized contribution in the z direction. The schematic on the right-hand side shows the corresponding density of states of the conduction band for a 2D quantum well. The band gaps of the bulk material and the quantum well are indicated in this figure: because of the quantification of energy levels due to confinement, the lowest energy level in the conduction band is energetically "higher" than the conduction band minimum in the bulk material, hence the bandgap of the well is larger. The DOS is derived in the text and summarized in Equation 1.7.

where m^* is the electron effective mass, and E is the energy. The de Broglie wavelength of the typical charge carrier for metals is usually only a few nanometers; the values for semiconductors are considerably larger, for example GaAs has a de Broglie wavelength of 24 nm[1], while Si has a smaller de Broglie wavelength of around 12 nm. Therefore a metal cluster, as discussed in Chapter 4, will only show confinement effects for very small nanometer-sized clusters, while quantum dots made of semiconductor materials can be an order of magnitude larger and will still exhibit the characteristic 0D DOS with discrete energy levels as a signature of effective confinement.

The experimental challenge now lies in the creation of specific confinement potentials, which provide one-, two-, and three-dimensional

[1] http://www.ioffe.ru/SVA/NSM/ is a database of essential semiconductor properties and bandstructures. The de Broglie wavelengths of several semiconductors are included here: http://www.ioffe.ru/SVA/NSM/Semicond/index.html.

confinement. In reality the height of the confining barrier is not infinite and usually reaches a few tenth to several electron volts at the most. The height of the barrier defines firstly, the temperature, at which charge begins to escape the well due to thermal excitation, and, secondly, the barrier height and distance between adjacent wells determines the tunneling probability and thus charge exchange through extended structures. This becomes important when we build superlattices where the interaction between nanostructures becomes a component of the overall functionality. A confinement potential can be created by interfacing heterostructures with dissimilar band gaps, which can be designed to trap electrons or holes within the well, or we can use the surface, the interface between solid and vacuum [7].

A beautiful example of electron confinement is the so-called quantum corral, where Crommie *et al.* [8, 9] showed for the first time the "image" of an electron wave or more precisely, the spatial variation of the DOS of a standing electron wave in a spherical, planar confinement potential. The quantum corral, which is shown in Figure 1.2 was built from 48 Fe atoms, which form a circle. The atoms are positioned by moving them with the tip in a scanning tunneling microscope, and the sample temperature is sufficiently low to minimize thermal motion of Fe ad-atoms. The electron wave is triggered by injection of an electron at the center of the quantum corral, and is confined by the "wall" of Fe atoms. Confinement is not

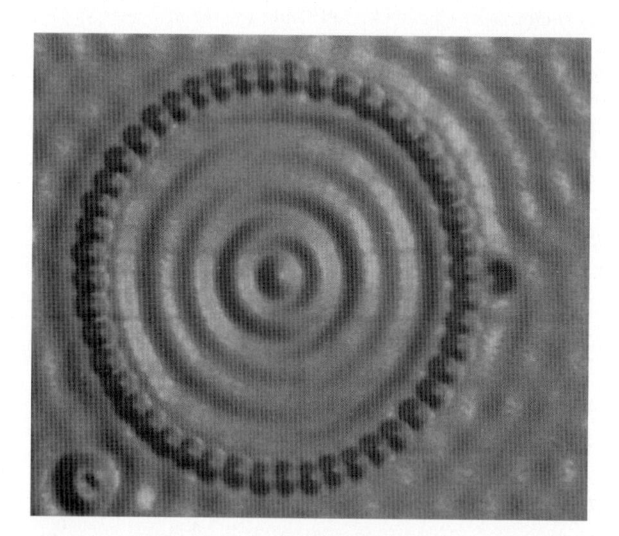

Figure 1.2 Quantum corral [22] of 48 Fe atoms on Cu(111). The Fe atoms are moved into position by the STM tip, which also serves to image the standing electron wave within the corral. The Fe-atom corral acts as confinement potential and the wavelength of electron wave agrees with the solutions of the Schrödinger equation for a planar system. From [22]. Reprinted with permission from AAAS. The individual steps in the quantum corral assembly are shown in some more detail at http://www.almaden.ibm.com/vis/stm/corral.html.

complete, and some small amount of charge leaks out of the corral, which can be seen in the image as a rapidly decaying wave on the outside of the corral. However, the standing wave inside the corral is truly two-dimensional: it is not only confined by the corral but also confined to the surface of the Cu crystal. This appears somewhat counterintuitive since Cu is a metal, but the Cu bulk band structure has a band gap in the (111) direction. The d-band of Cu is positioned a few eV below the Fermi energy, and the sp-band crosses the Fermi energy in the (110) and the (100), but not in the (111) direction [10]. The 3D band structure of Cu therefore has no electronic states in this direction, but the 2D band structure of the Cu(111) surface presents so-called surface states, which can be calculated from the solution of the Schrödinger equation for the 2D surface lattice. The band gap in the bulk, 3D band structure confines this 2D state to the surface, and prevents propagation of the electron wave into the bulk. The confinement of the surface state, and the confinement by the artificial corral made of Fe atoms are required to produce the beautiful standing wave, which can then be imaged with scanning tunneling microscopy (STM).

1.1.1 Density of States for 3,2,1,0 Dimensions

The reduction of dimensionality changes the electronic structure decisively and we derive here the density of states (DOS) for the 3D to 0D case. The DOS $\mathrm{d}N/\mathrm{d}E \cdot 1/V$ of an electronic system is defined as the number of electronic states $\mathrm{d}N$, which occupy a given energy interval $\mathrm{d}E$ per unit volume (V). The DOS for electrons within a 3D solid can be derived as following, starting from the expression for the energy E_n of a state within the solid as a function of the wave vector k. The electrons are in this case treated as free electrons, and the spatial variation in the Coulomb interaction with the ion cores is neglected. The energy is given for the 3D case in analogy to Equation 1.1 by:

$$E_n = \frac{\hbar^2 \vec{k}^2}{2m^*} = \frac{\hbar^2}{2m^*}(k_x^2 + k_y^2 + k_z^2) = \frac{2\pi^2\hbar^2}{L^2 m^*}(n_x^2 + n_y^2 + n_z^2) \qquad (1.3)$$

with the wave vector component $k = 2\pi n/L$ (each wave vector is described here as a cube whose position in space is determined by the associated quantum numbers, whereas L is the side length of a cube of solid material). n_x, n_y, and n_z are the quantum numbers which characterize each individual energy state, which can be occupied by two electrons of opposite spin. We now use a geometrical description of the space of quantum numbers, and redefine n_x, n_y, and n_z as a position vector (n_x, n_y, n_z). The endpoint of each of these vectors is a unique position in space, which corresponds to exactly one energy state and a primitive unit cell, which can hold two electrons of opposite spin. The volume of an allowed state in k space is then given by $(2\pi/L)^3$. The highest occupied energy state at 0 K is the Fermi energy, which corresponds in this geometric picture to a sphere. The DOS is obtained by dividing the volume of a spherical shell (which corresponds to an energy interval) by the volume of a single state. The volume of the spherical shell from radius $\vec{k}$ to $\vec{k} + d\vec{k}$ is given by:

$$V_{shell}\mathrm{d}\vec{k} = 4\pi|\vec{k}^2|\mathrm{d}\vec{k} \tag{1.4}$$

and is then divided by the volume of the single state, which yields the number of states in this shell:

$$\frac{\mathrm{d}N}{\mathrm{d}\vec{k}} = \frac{L^3|\vec{k}^2|}{\pi^2} \tag{1.5}$$

Using the relation between energy and k vector introduced in Equation 1.3, we obtain $\mathrm{d}E/\mathrm{d}k$ and can now determine the number of states per energy interval for a 3D system of free electrons:

$$\mathrm{DOS}_{3\mathrm{D}} = \frac{\mathrm{d}N_{3\mathrm{D}}}{\mathrm{d}E} - \frac{1}{L^3} = \frac{1}{2\pi}\left(\frac{2m^*}{\hbar^2}\right)^{3/2} E^{1/2} \tag{1.6}$$

The expressions for the ideal 2D and 1D systems can be derived in the same way as for the 3D case, and Figure 1.3 schematically shows the DOS for 3D to 0D systems. For the 2D system (e.g., a quantum well) one component in k space is fixed; for the 1D system (e.g., a nanowire) two components in k space are fixed. In the 2D system each state occupies an area of $(2\pi/L)^2$, and in the 1D system this is reduced to a length of $(2\pi/L)$. The derivation of the DOS for the 2D and 1D system is then made in analogy to the 3D case, using the area of an annulus (2D) and the length of line segment (1D), respectively.

The $\mathrm{DOS}_{2\mathrm{D}}$ is independent of the energy, and becomes a step function:

$$\mathrm{DOS}_{2\mathrm{D}} = \frac{m^*}{\pi\hbar^2} \tag{1.7}$$

And the DOS for the 1D system:

$$\mathrm{DOS}_{1\mathrm{D}} = \frac{1}{\pi}\left(\frac{2m}{\hbar^2}\right)^{1/2} \frac{1}{E^{1/2}} \tag{1.8}$$

The DOS, which are derived here, are the solutions for the ideal system, for example, a 2D sheet without any contributions from a third direction. For a 0D system such as a quantum dot, the energy levels become fully discrete, and the DOS shows only discrete lines, which correspond to the

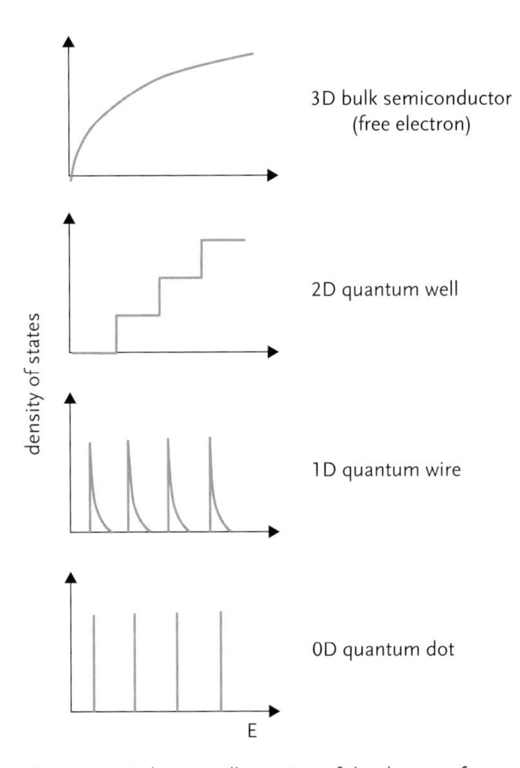

Figure 1.3 Schematic illustration of the density of states for 3D to 0D systems.

atom-like discrete spectrum. Semiconductor quantum dots or ultrasmall metal clusters are therefore often labeled as "artificial atoms".

1.2
Synthesis of Nanostructures: Fundamental Surface Processes and Reactions

The rapid change in properties and stability as a function of size and dimensionality places rather stringent conditions on the synthesis of nanostructures. A multitude of different methods has been developed, and we often distinguish between top-down and bottom-up approaches. Lithography-based techniques, which are essential in semiconductor device fabrication, are top-down techniques, where an artificial structure is constructed by sequential deposition and etching steps. The length scale is defined and limited by the wavelength and energy of the electron, ion, or photon beam, which is used to write the respective structural features.

In a bottom-up approach the nanostructure is essentially assembled from the smallest unit, often aided by the use of a template or a self-assembly process. This approach allows access to the size regime below 10 nm, which can currently not be reached by lithography. The methods used for the synthesis of the basic nanoscale units, such as the semiconductor nanowires, carbon nanotubes, or quantum dots, are highly diverse and are discussed in detail in the chapters devoted to these groups of materials. In this introductory chapter we will provide the reader with a basic understanding of the fundamental surface processes and reactions.

The importance of fundamental surface processes for controlling shape, structure, and composition was recognized early on in systematic studies of crystal growth. Understanding the interplay between surface reactions has become even more critical in order to achieve control of nanostructure synthesis and to reach at least some degree of predictability in designing new nano-sized building blocks and arrays of building blocks. The fundamental surface processes can be subdivided into four groups, as shown schematically in Figure 1.4: (i) adsorption and desorption, (ii) diffusion, (iii) nucleation, and (iv) growth. While all of these processes are intimately linked, it is helpful to discuss them separately, at least in the beginning.

However, before we start to discuss the fundamental surface processes we must take a closer look at the electronic and geometric structure of a surface. In the so-called terrace-step-kink (TSK) model each atom is represented by a single cube. This representation of the surface was also chosen in Figure 1.5. The TSK model helps to recognize different geometric defects on the surface, such as step edges, point defects, and kink sites, which are under-coordinated lattice atoms and thus often serve as preferential reaction or nucleation sites. The TSK model, however, fails to represent the geometric structure on the atomic scale. If a solid is cleaved along a certain lattice plane, the atoms, which are now positioned on the new surface, experience a dramatically different bonding environment, and consequently a reorganization of the geometric structure is often necessary

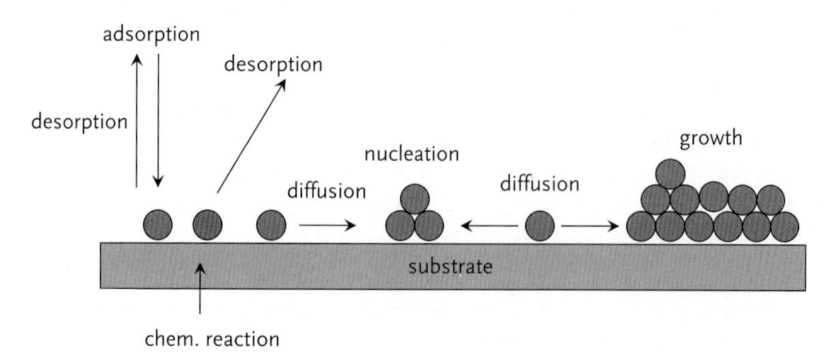

Figure 1.4 Schematic illustration of surface processes in thin film and nanostructure growth including adsorption–desorption, diffusion, nucleation, and growth.

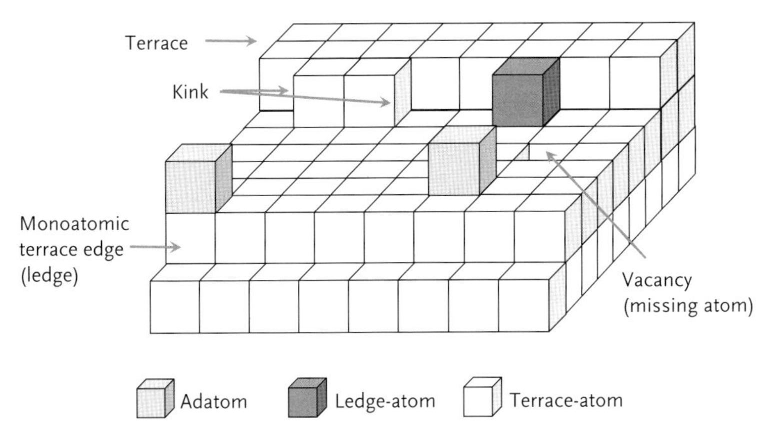

Figure 1.5 A model surface (TSK model) to illustrate the coordination number of different atomic bonding sites on a surface.

to reach a new equilibrium state. The asymmetry in the electron screening at the surface plane leads to a relaxation of the outermost atomic layers, and periodic modulation of the interatomic distances perpendicular to the plane of the surface, but leaves the in-plane atomic positions intact; the symmetry of the lattice plane is conserved. This relaxation is most often observed for metals, while covalently bonded materials usually undergo a so-called reconstruction, where the atomic positions within the plane of the surface are modified and deviate from those of the equivalent plane within the 3D solid. The surface reconstruction is treated thermodynamically and kinetically like the formation of a new phase, and several different reconstructions can co-exist on a single surface orientation. The surface reconstructions can offer unique templates with atomic-scale corrugation and highly selective attachment sites. The guided or templated self-assembly of nanostructures on reconstructed surfaces has been used extensively, and several examples are included in Chapters 3 and 4.

The growth of a nanostructure or thin film is initiated by the adsorption of ad-atoms as the first step. When looking at an adsorption process we often distinguish between physisorption and chemisorption. Physisorption is characterized by weak bonding between surface and adsorbate. In Figure 1.6 the interaction potential for a fairly common reaction, dissociative chemisorption, is shown as a function of distance. The molecule A_2 approaches the surface, is physisorbed, and enters the shallow physisorption well. The weak, physisorption interaction is due to van der Waals forces, which arise from the interaction between dipoles created by quantum mechanical fluctuations of the electron charge distribution. The interaction between the incoming adsorbate atom/molecule and the electrons of the solid can be described reasonably well as a harmonic oscillator, where the oscillator coordinate is the trajectory of the incoming atom

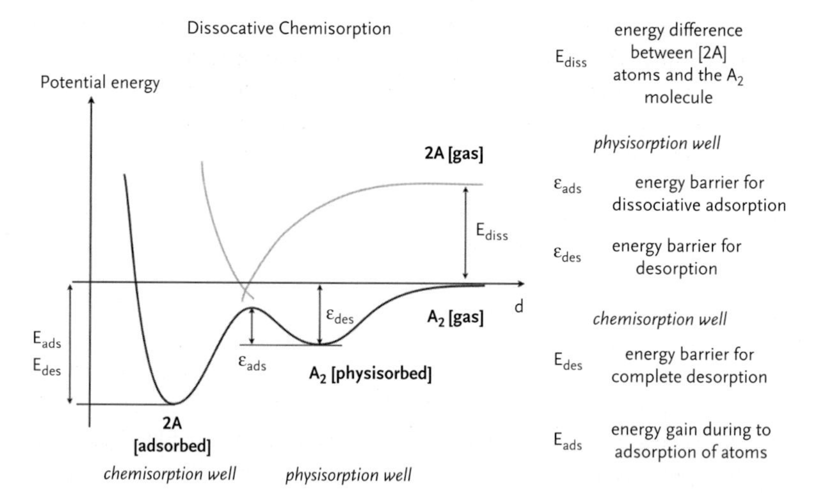

Figure 1.6 Characteristic potential well of the relatively common dissociative chemisorption; the energy is shown as a function of the surface–ad-atom separation, and the origin of the energy axis is the potential energy of a molecule far away from the surface. A wider range of surface reactions and interaction potentials is discussed in [13].

normal to the surface [11]. The adsorption of noble gases at surfaces only shows a physisorption well, and the energy levels within that well agree closely with the simple harmonic oscillator model.

The repulsive term at shorter distances comes from the Pauli exclusion principle, due to the onset of overlap between the closed shell electrons in the atom and the electrons "spilling out" of the solid. The equilibrium distance of a physisorbed atom or molecule from the surface is relatively large, and about 0.3 nm above a noble metal surface for He [12]. The minimum in the potential well is shallow and the physisorption binding energy consequently relatively small. Energy levels within a physisorption well have been probed by scattering experiments. For comparison, the bond length for a covalent bond is only about 0.1–0.15 nm. Each atom or molecule approaching a surface will move through the physisorption potential well before the much stronger, short range chemisorption interaction begins to dominate, or the atom is ejected back into the vacuum. The ad-atom can only be captured in an adsorption well if it can reduce its kinetic energy sufficiently to occupy one of the energy states within the well. This relaxation is often achieved by electronic or vibrational excitation of the surface, which is then dissipated into the bulk. In the example for dissociative chemisorption the activation energy to the chemisorption well with two A atoms is lower than desorption into the vacuum or adsorption as a A_2. Dissociation is therefore the favored reaction pathway.

The chemical interaction of an adsorbate with a surface is considerably more complex than physisorption. A reasonable description of physisorption can be achieved by treating the surface as homogenous and featureless entity; however, the formation of a chemical bond is highly sensitive to the local geometric and electronic structure of the surface. A multitude of reaction scenarios is known, and have been investigated in great detail in the study of surface reactions of relevance in catalysis. The chemisorption process is particularly complex for molecules where the orientation of the molecule relative to the surface, the surface structure, and rotational and vibrational internal degrees of freedom within the molecule determine the reaction pathway. The reaction process between ad-atom and surface can be described by using a potential energy surface with two (three) coordinates: one (two) in the surface, and one for the distance between ad-atom and surface. This description captures the relative orientation of the ad-atom with respect to the surface, and includes the physi- and chemisorption wells.

An example of the interplay between adsorption and desorption processes in the synthesis of nanostructures is the formation of graphene sheets: one method uses the pyrolysis of hydrocarbon molecules on transition metal surfaces, and another frequently used approach employs desorption of silicon from a SiC surface, so that the graphene layer is left behind. These processes are described in detail in Chapter 6.

Adsorption and desorption processes are the initial steps in the synthesis of thin films and nanostructures, and the ad-atom bonding site is determined by the random selection of the initial impact site and the shape of the potential well as a function of position and ad-atom orientation. However, the initial bonding site might only present a local minimum in the potential energy surface. Probing of adjacent sites and motion into a lower energy position can only be accomplished by diffusion. Diffusion can happen on a very short length scale and just be perceived as a repositioning of the ad-atom, or it can involve transport on long length scales across a large section of the surface. The interplay between the kinetic aspects, such as diffusion, and the thermodynamic drive to minimize the energy of the system can be used to modulate nanostructure geometry over a very wide range. Diffusion of ad-atoms involves a multitude of different mechanisms on an atomic length scale, and a detailed discussion can be found in many surface science textbooks, such as Ref. [13]. We are concerned here mostly with the impact of diffusive and kinetic processes on the growth, morphology, and geometry of nanostructures.

The homoepitaxy of metal films is a particularly powerful example to illustrate the interplay between thermodynamic and kinetic control of surface processes. The homoepitaxy of several metals is very well studied and an extensive discussion is given in Ref. [14]. In thermodynamic equilibrium the deposition of Pt on a Pt surface should lead to a layer-by-layer growth; no strain is built up at the interface and a Pt overlayer perfectly wets a Pt surface [14–17]. However, the layer-by-layer growth and consequently formation of a smooth Pt overlayer is only observed at relatively high temperatures, which indicates the presence of energy barriers, which inhibit the movement of Pt ad-atoms into energetically favorable bonding sites. The deviation of the homoepitaxy of Pt on Pt from the layer-by-layer growth mode can only be explained by kinetic arguments, and the presence of an activation barrier of sufficient height to impede the adoption of the energetically most favorable layer structure. We will take a closer look at two of the wide range of structures which form in the homoepitaxial growth of Pt layers and are shown in Figure 1.7: the fractal structures, which form at relatively low temperatures and are most prominent at low coverages, and the pancake, or stacked, layer structure, whose formation requires an elevated temperature.

The fractal structure forms in a so-called "hit and stick" process: the Pt ad-atoms possess only limited mobility on the surface, and will become immobile once they bond with another highly unsaturated ad-atom. They cannot overcome the activation barrier for detachment from an energetically unfavorable position, and will "stick" at their initial attachment site after they "hit" another ad-atom or an unsaturated bond. Many fractal structures show a preferential orientation with respect to the underlying surface structure, which is an indication of preferential, easier directions of

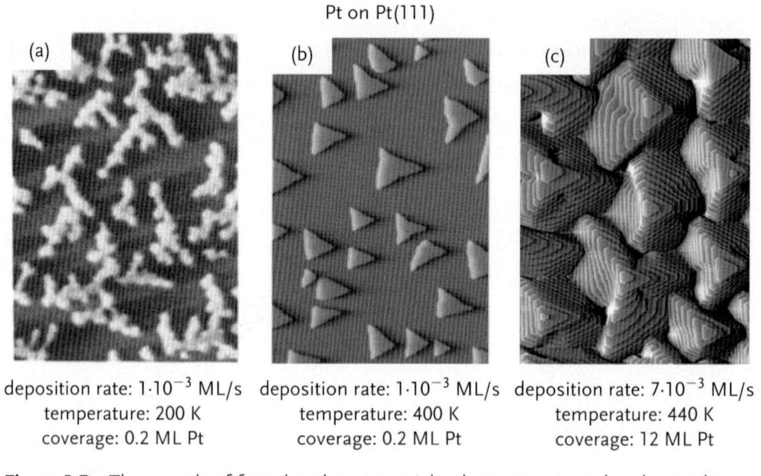

Figure 1.7 The growth of fractal and compact island structures can be observed in the homoepitaxy of Pt on the Pt(111) surface [15–17]. The structural variability is controlled by a hierarchy of activation barriers. The critical processing parameters are deposition rate, temperature, and coverage and they control structure formation during homoepitaxial growth on a clean sample. The presence of traces of some adsorbates such as CO can alter the activation barriers decisively and push the system into a different growth mode. Fig. 1.7(a) and (b) reprinted with permission from [15] and [16] . Copyright (1998 and 1993, respectively) by the American Physical Society. Fig. 1.7 (c) is reprinted from [17] with permission from Elsevier.

diffusion (diminished activation barrier for diffusion) or detachment. Fractal structures have also been observed for fullerene islands on a fullerene surface [18, 19], and are present in many low temperature processes. If the temperature is increased many of the ad-atoms, which are initially positioned in unfavorable bonding sites can now reposition themselves and probe a multitude of sites.

The critical barrier for ad-atom motion, which leads to the formation of stacked structures, is the Ehrlich–Schwöbel barrier [20, 21]. This is the additional energy required for motion of an ad-atom across a step edge from the higher to the lower terrace compared with the activation barrier to move on the terrace. The origin of the additional barrier is a rearrangement in the spatial distribution of electrons around a highly unsaturated step edge atoms. Any ad-atom landing on the upper level of a terrace is now trapped on that terrace, and will be repulsed from the step edge. The distance between step edge and the boundary of the new island on the upper level corresponds to the diffusion length of the ad-atom. The magnitude of the Ehrlich–Schwöbel barrier is strongly system and material dependent. Once the temperature is sufficiently high, and ad-atoms can overcome this activation barrier, the layer-by-layer growth, which is predicted based on solely thermodynamic arguments, will proceed. The interplay between a multitude of activation barriers is key to achieving a deterministic approach in nanostructure fabrication, and a kinetic description based on the individual reaction rates for the atomistic processes has to be used for a full comprehension of growth far from equilibrium.

The next step in the growth of layers, wires, and nanostructures is the formation of a stable nucleus, which then proceeds to grow by further accumulation of ad-atoms. Classic nucleation theory as described in many textbooks has many limitations, but it is used here to illustrate the role of the surface energy during the initial nucleation phase of a seed crystal. The condensation of the solid seed crystal is driven by the change in Gibbs free energy with nucleus size – the total change in Gibbs free energy is the sum of bulk, surface and interface contributions. The contributions to the change in Gibbs free energy as a function of nucleus radius r are a volume component given by:

$$\Delta G_{\text{bulk}} = \frac{4}{3}\pi \rho r^3 \Delta \boldsymbol{\mu}_{s,l} \tag{1.9}$$

and a surface contribution

$$\Delta G_{\text{surface}} = 4\pi r^2 \gamma_{s,l} \tag{1.10}$$

where ρ is the density of the solid, $\Delta\mu_{\text{s,l}}$ is the chemical potential difference, and $\Delta\gamma_{\text{s,l}}$ is the interfacial energy at the solid–liquid s,l interface. The surface contribution to the Gibbs free energy dominates for small particles prior to reaching the critical nucleus size, and can indeed drive the formation of a phase, which is surface stabilized and does not represent the most stable bulk structure. At larger particle sizes, the bulk contribution becomes decisive due to its r^3 dependence on the particle radius. The surface-stabilized nucleus will only survive if the energy difference between the competing phases is relatively small, which is indeed the case for many of nanowires made of compound semiconductors considered in Chapter 3.

1.3 Closing Remarks

Confinement and dimensionality are concepts which are at the center of nanoscience and technology and will accompany us throughout this book. A nanoscale structure and material is defined with these concepts in mind, and they are the driving force for the rapid and often astonishing modification of properties as a function of diminishing size. In this book we mostly focus on electronic, and in some instances optical and chemical, properties of nanoscale materials, and confinement is therefore defined in terms of characteristic length scales inherent in the electronic structure, such as de Broglie wavelength.

2
Experimental Techniques for Nanoscale Materials Analysis

The investigation of nanoscale materials requires analytical techniques which are suitable to study the structure and functionality of nanoscale materials and building blocks. The wide range of analytical techniques which are at our disposal, can be broadly divided into active techniques, where the nanostructure is incorporated within a specific device configuration, and passive techniques, which probe nanostructures in an environment not related to its potential application. Electrical transport measurements are an example of an active measurement; the nanostructure becomes an integral part of a device, and its response to the application of an electric field is recorded. The active techniques are highly versatile, and will be discussed as required in the chapters on individual materials; one specific example is the single electron transistor, which is described in detail in Chapter 7.

Passive analytical techniques include microscopy techniques, such as scanning probe microscopy (SPM) and transmission electron microscopy (TEM), and spectroscopy techniques, such as electron spectroscopy, optical methods, Raman spectroscopy, and magnetic probes. However, some of these analytical tools can also be used for the synthesis and manipulation of nanoscale structures. The manipulation of surface atoms to create a quantum corral with a scanning tunneling microscope [8, 9, 23] (STM), an atomic scale computer [24], and the writing of arbitrary nanoscale pores with the electron beam in a transmission microscope [25] are just a few of many examples for nanoscale manipulation with analytical tools. Scanning probe techniques have been instrumental in the development of nanoscience, and therefore play a prominent role in this book.

The present chapter focuses on the passive techniques, and gives an overview of SPM, and electron spectroscopies, which are used in examples throughout this book. These techniques are often new to students who are just embarking on the exploration of nanoscience and this chapter is therefore intended as an introduction to the techniques.

Scanning probe microscopy (SPM) is based on two closely related techniques, which can offer atomic resolution: scanning tunneling microscopy (STM), where the image contrast is derived from the modulation of

Inorganic Nanostructures: Properties and Characterization, First Edition. Reinke, P.

Published 2012 by WILEY-VCH Verlag GmbH & Co. KGaA

the quantum mechanical tunneling current through the tip–sample interaction, and atomic force microscopy (AFM), where the interaction force between a tip and the sample surface is measured. Historically, the invention of STM and AFM lead to a rapid rise in nanoscience research since they offered, for the first time, a view of the atomic level processes and properties for a wide range of materials. A few decades earlier the field ion microscope (FIM) had been the first instrument to show images of metal surfaces with atomic resolution [26] but it lacked the versatility of STM and AFM, and therefore never acquired the widespread popularity of SPM methods.

2.1
Scanning Probe Microscopy

2.1.1
Scanning Tunneling Microscopy – STM

In scanning tunneling microscopy the image contrast is provided by the modulation of the quantum mechanical tunneling current, which flows between sample surface and tip. The electron tunneling through a vacuum barrier between tip and sample was discussed controversially in the early days of STM, but it has since then been shown that the quantum mechanical description of electron transmission through a potential barrier is equally applicable to the understanding of the tunneling process in an STM [27, 28]. The dimensions of the tunneling barrier between tip and sample surface is defined by the respective work functions and bias voltage between tip and sample (barrier height), and tip–sample distance (barrier width). The atomic resolution, which is routinely achieved in STM experiments, is due to the exponential, rapid decay of the current with tip–sample distance and concomitant sensitivity to small changes in the tip–surface interaction.

Figure 2.1 illustrates an STM measurement: a bias voltage U_B is applied between tip and sample, and controls the relative positions of the respective Fermi levels, E_{F1} and E_{F2}. The tunneling current then flows (in this figure) from the occupied states in the tip to the unoccupied states of the sample thus probing the sample's empty states. For simplicity the sample is assumed to be a metal. The filled states of the sample will be probed if the bias voltage is inverted. In theoretical descriptions of the tunneling process it is generally assumed that the tip itself does not change during the measurement, and that the density of states (DOS) at the apex of the tip is spherical [29–31].

However, in reality the exact tip shape is usually not known, and ideally only the last atom at the tip end participates in the tunneling process. The sharpness of the tip end can only be deduced indirectly from the quality of the image, or by imaging well-known surface structures such as the

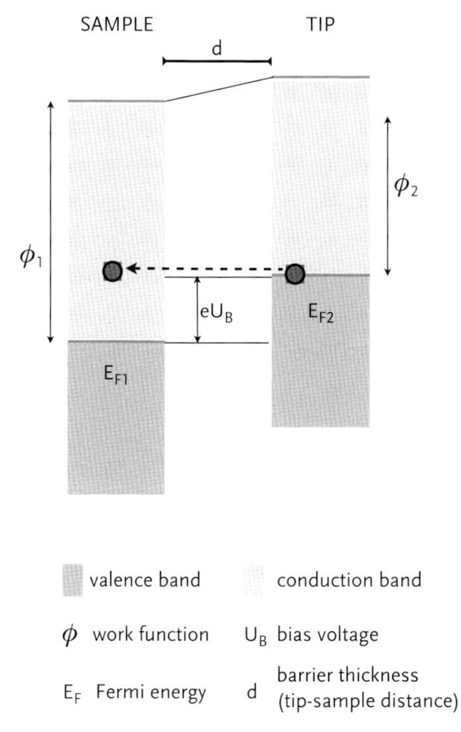

Figure 2.1 Schematic depiction of the electron tunneling process during STM imaging. The sample and tip valence and conduction bands are shown, and the bias voltage applied between them is expressed in the shift of the Fermi energies. The barrier height is determined by the respective work function, and the barrier width is the distance d between tip and sample. The bias voltage controls which states contribute to the tunneling current. This example shows tunneling from the filled states of the tip to the empty states of the sample, inversion of the voltage leads to tunneling from filled states of the sample to empty states of the tip.

Au-herringbone [32] reconstruction or the Si(111)-(7 × 7) surface [33, 34]. Modification of the tip during a measurement is an experimental challenge, and rapid changes in the image quality are frequently observed due to modulation of the tip shape or attachment of ad-atoms or molecules.

The magnitude of the quantum mechanical tunneling current [29–31] I_{s-t} ($s =$ sample, $t =$ tip) depends on the local DOS and the distance between tip and sample. Eq. 2.1 follows the model developed by Tersoff and Hamann [30], and fully describes the quantum mechanical tunneling process in the limit of small voltages:

$$I_{s-t} = \frac{4\pi e}{\hbar} \int_0^{eV} \rho_s(E_F - eV + \varepsilon)\rho_t(E_F + \varepsilon) M^2 \, d\varepsilon \tag{2.1}$$

where (i) E_F is the Fermi energy; (ii) M is the quantum mechanical transition matrix element, which describes the overlap of the tip and sample wave functions; (iii) ρ_s and ρ_t are the DOS of the surface and the tip, respectively; (iv) ε is the energy of the energy level participating in the tunneling process with respect to the Fermi energy; and (v) eV is the bias voltage. The electrons tunnel from states close to the Fermi energy in the tip $\rho_t(E_F+\varepsilon)$ to empty states in the sample $\rho_s(E_F-eV+\varepsilon)$ in the case described here. The tunneling current is the integral current with contributions from all states positioned between the bias voltage and E_F. The derivative dI/dV is the representative of the DOS itself. To describe tunneling from sample to the empty states in the tip the indices s and t are reversed. This expression, and specifically the transition matrix element, can be simplified by assuming a flat DOS and a spherical shape of the tip wave function.

The tunneling current depends on the overlap of the electron wave functions of tip and sample, which defines the transmission probability. The transmission probability contains information on the local DOS, and the localization of the sample's electron wave function above the surface plane. Tunneling from states or orbitals which have only a small extension into the vacuum gap such as d-orbitals/states can lead to a smaller tunneling current than for example a p-orbital, whose wave function often extends much farther into the vacuum gap. These examples are discussed in detail by Tromp in an extensive review about scanning tunneling spectroscopy [35].

Any image recorded in STM therefore contains information on surface topography and surface electronic structures, and this is illustrated quite well in Figure 2.2. This figure shows a small section of the surface of a polycrystalline vanadium dioxide surface [36], and the images are recorded with bias voltage of 0.2 V. VO_2 undergoes a phase transition from an insulator to a metal at 341 K, and the left hand image is taken below the transition temperature, and the right hand image is recorded above the transition where the surface is metallic with a considerably higher DOS within the energy interval between 0.2 V and E_F. The cluster of crystallites appears considerably brighter for the metallic surface, and the increase in apparent height is solely due to a modification of the DOS and not a change in the surface topography.

STM images can be acquired in constant current and constant height modes, the latter is mostly used for scanning tunneling spectroscopy (STS) measurements. In constant current mode, which is often called topography mode, the tunneling current is kept constant by use of an electronic feedback loop. The feedback loop reacts to changes in the tunneling current and adjusts the tip height accordingly to keep the feedback current at a preset value. However, the tip-height modulation occurs not only in response to topography changes, but also due to differences in the local DOS (LDOS) as illustrated in the previous paragraph.

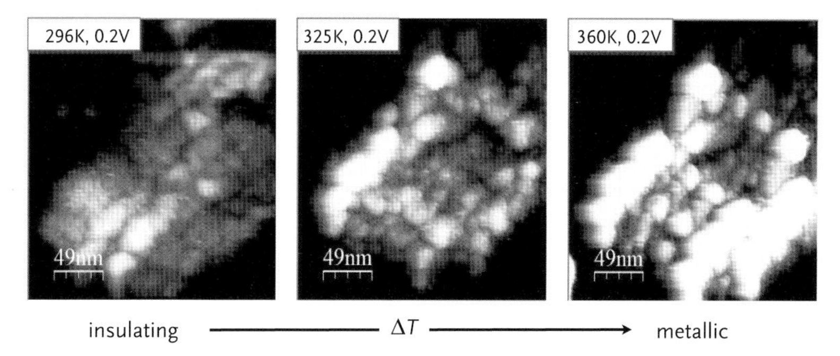

Figure 2.2 STM image of a vanadium dioxide thin film [36]. VO_2 undergoes a phase transition at 341 K, which transforms the insulating monoclinic phase to the tetragonal metallic phase. The image is recorded with a bias voltage of 0.2 V at different temperatures. This voltage probes within the band gap of the insulating phase, and the tunneling current is carried by the relatively few defect states at the surface. For the metallic phase, which dominates at the highest temperature, the current is considerably higher due to the population of states at the Fermi energy. The apparent height modulation is driven by the electronic density of states and not by a change in topography. Reprinted with permission from Journal of Applied Physics [36], Copyright 2010 by American Institute of Physics.

The sensitivity of the tunneling current to the LDOS (local density of states) at the same time gives us an unparalleled tool to investigate the electronic structure of surfaces and nanostructures. In STS the bias voltage is varied, while the feedback loop is switched off, and thus the tip–sample distance is kept constant. The derivative of the current–voltage characteristics, the differential conductance dI/dU, is related to the LDOS. The normalized differential conductance is obtained by dividing the differential conductance by the total conductance I/U, and is given by $(dI/dU)/(I/U) = (d \ln I)/(d \ln U)$. The normalization of the differential conductance minimizes the influence of the dependence of the tunneling current on tip–sample distance as shown by Lang [37]. STS therefore measures the electronic structure of a surface, and offers an unprecedented view of atomic bonding.

A recent example for the application of tunneling spectroscopy in the characterization of an inorganic low-dimensional structure is the measurement of the distribution of charge puddles (local areas with accumulation of a single kind of charge leading to substantial spatial inhomogeneity) in graphene layers. Graphene is discussed in more detail in Chapter 6 and is a highly coveted material due to its ultra-high charge carrier mobilities. However, the measured charge carrier mobilities are often considerably below the expected, theoretical value. The most striking differences in the mobility values are found between graphene layers on SiC substrates and free-standing graphene membranes.

Figure 2.3 shows an example of a differential conductance map dI/dV of the graphene surface for different bias voltages. The dI/dV maps are

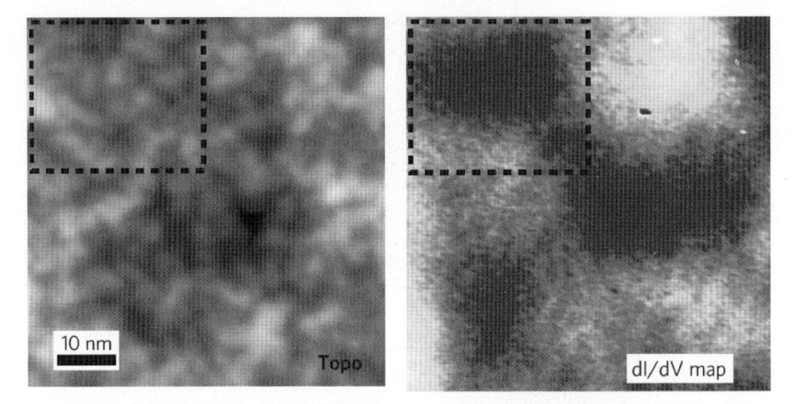

Figure 2.3 STM images of graphene [48]: (a) topography image recorded in the constant current mode with a bias voltage of -0.225 V (tunneling into filled states of sample) and feedback current of 20 pA, (b) is the dI/dV map of the same area ($V_{(bias}=0.225$ V, $I=20$ pA and gate voltage of 15 V). The dI/dV map shows the local charge fluctuations, which are related to the shift of the Fermi energy with respect to the Dirac point (see Chapter 6 for details on the electronic structure of graphene). Variation in brightness in the topography images are predominantly from the sample roughness, the variation in brightness and thus contrast on the dI/dV map are dominated by the local charge distribution. The graphene flakes are positioned on an SiO_2 surface, and the substrate also serves as a backgate; two electrodes were positioned on the flake for the completion of the circuit. This figure is reprinted by permission from Macmillan Publishers Ltd. from Nature Physics [48], copyright 2009.

composed of the differential conductance for all points of measurement in an STM image at a given bias voltage close to the Dirac point (see chapter 6.4 for details). The dI/dV maps can be used to identify charge accumulation and local variations in the surface electronic structure. This study was able to relate topographic defects, substrate defects, and charge injection from the SiC substrate beneath the graphene sheet to the presence and localization of the charge puddles. The defects and related charge puddles are potent scattering centers, and are critical limiters for achieving the desired charge carrier mobilities.

Another example, which is discussed in detail in Chapter 3, illustrates the application of STS in the analysis of the electronic structure of nano-structures, is the measurement of the band gap as a function of diameter in Si nanowires. The Si nanowires were grown by a chemical transport method from silicon oxide powder, and it was thus possible to grow wires with diameters down to nearly 1 nm. It had been predicted that quantum confinement leads to a dramatic increase in the band gap for Si nanowires with a diameter of less than 3 nm. STS of hydrogen terminated Si nanowires showed clearly the modulation of the band gap as a function of wire diameter. The agreement with theory was excellent, and the onset of quantum confinement around 3 nm of nanowire diameter was confirmed.

2.1.2
Atomic Force Microscopy – AFM

In atomic force microscopy (AFM) the image contrast has its origin in the interaction forces between the tip and the sample surface. AFM is a highly versatile method and, in contrast to STM, is not limited to the study of conducting materials. The tip–sample interaction can be described by the well-known force–distance curves. The force–distance curve also included in Figure 2.4 represents the sum over all forces within the system: long range, attractive forces such as van der Waals, electrostatic, and capillary forces at large tip–sample separation, and an increasing contribution from short range, repulsive forces due to wavefunction overlap as the distance is decreased.

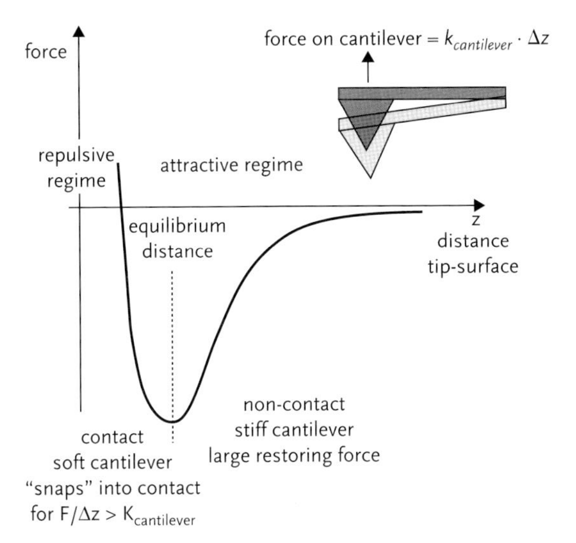

Figure 2.4 Schematic graph of the force distance curve between AFM tip and the surface. A force is exerted on the cantilever as it approaches the surface and initiates a bending of the cantilever, whose magnitude depends on its spring constant ($k_{cantilever}$). The repulsive regime is reached for positive forces, the attractive regime for negative net forces. Once a tip snaps into "contact" it will be positioned on the left hand side of the equilibrium distance.

An excellent example for comparison of STM and AFM is the imaging of the graphite surface with atomic resolution which can be achieved with STM and AFM, and a comparison of both images illustrates in a very powerful manner the differences in contrast mechanisms [49]. The low-temperature images of a graphite surface imaged with AFM and STM are shown in Figure 2.5. STM records the differences in the DOS around the Fermi level, and therefore only every second carbon atom, which does not have a direct bonding partner in the second layer (and no appreciable extension of the wavefunction into vacuum), will contribute to the image. AFM on the other hand is sensitive to the force modulation, which is driven by the overall charge density and therefore all atoms can be resolved.

The modulation of the interaction between tip and sample surface during the measurement is the origin of contrast in AFM measurements. The detection of these modulations is in the majority of instruments achieved by measuring the bending of the cantilever (the tip is attached to one end of

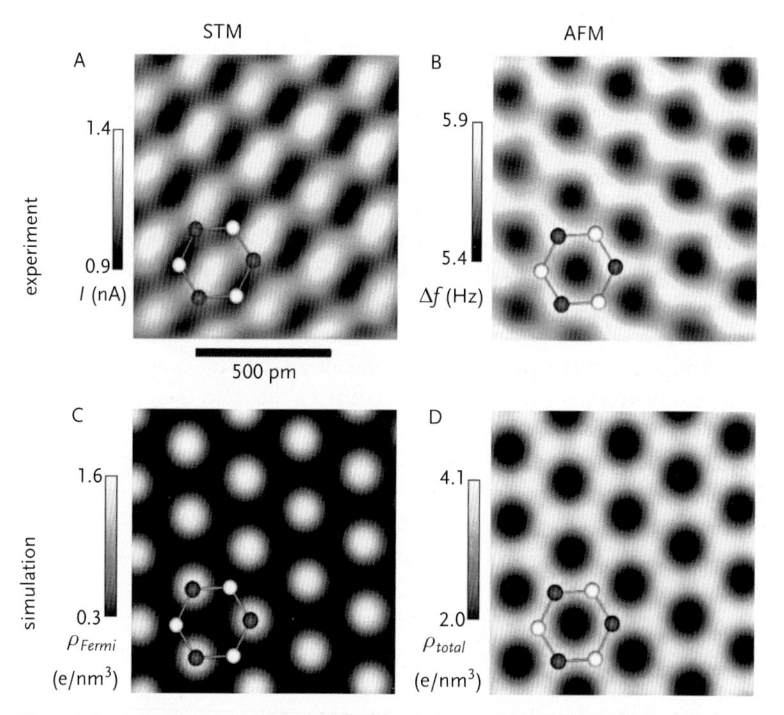

Figure 2.5 Images of a graphite surface recorded with STM and AFM [39, 49]; the experimental images are compared with simulated images. In the STM images of graphite only every second atom is detected, and the carbon atoms with a direct bonding partner in the second layer will not contribute to the image due to a lack of localization of charge at the surface (the unit cell of graphite is described in Chapter 6). In AFM both atoms can be detected, since the contrast is driven by the interaction forces. These images were taken at 5 K in a custom-built instrument which allowed for simultaneous AFM and STM measurements. Figure from [49] is reprinted with permission from Proceedings of the National Academy of Science (PNAS), copyright 2003, U.S.A.

the cantilever bar), which is proportional to the forces acting on the tip [38]. This is schematically illustrated in Figure 2.4. The cantilever's response can be described as that of a spring, whose spring constant is determined by the material and dimensions of the cantilever bar. Several methods have been developed to measure the response of the cantilever to the forces acting on the tip, and they are often subdivided into static and dynamic methods (we follow here the nomenclature as presented in Ref. [38]). In the static mode the deflection of the cantilever is measured directly, for example, by sensing the motion of a laser beam reflected from the cantilever onto a four-quadrant detector [38]. Methods which use a tuning fork integrated into the cantilever support have recently gained a larger market share, and allow to switch rapidly between operation in STM and AFM modes without tip exchange.

The most frequently used measurement mode in conjunction with a static measurement of the cantilever bending is the so-called contact mode. The tip moves along the force–distance curve as it approaches the surface and "snaps" into contact once the tip–surface interaction overwhelms the restoring force of the cantilever spring. On the force-distance curve the z-position of the tip is now on the left hand side of the force-distance minimum. The tip is then positioned very close and "in contact" with the surface, the forces acting on the sample are substantial and can even be destructive to the surface.

In the dynamic mode of operation, which is often used in conjunction with non-contact methods, the cantilever vibrates with its characteristic mechanical eigenfrequency and the frequency shift due the forces acting on the cantilever is detected. The cantilevers used for non-contact measurements are usually rather stiff, while the contact mode employs softer cantilevers with a much smaller spring constant and thus restoring force. The operation of this type of dynamic non-contact measurement often requires a vacuum environment to achieve the sufficient stability in the oscillation frequency of the cantilever. This is not the case for all types of non-contact measurements, and a variety of different modes has been developed for measurements in air or liquids. The tapping mode, which is used with great success in non-vacuum and liquid environments, relies on an intermittent contact between tip and sample surface. The amplitude of the tip oscillations is substantial (>10 nm range) and the amplitude variation is measured at a constant frequency close to the mechanical eigenfrequency of the system; the tapping mode is classified as a dynamic mode measurement. Numerous techniques, which combine contact, non-contact, dynamic, and static modes have been developed over the years, and we have now the luxury to choose the most suitable method for a wide range of experimental questions.

Unfortunately it is not as easy to obtain atomic resolution as the many beautiful images in the literature suggest. Even if we can assume that the tip is an "ideal" tip where only a single atom protrudes from the tip end, this tip–atom will interact with more than a single substrate atom. The extent of the interaction of the tip with the next nearest surface atom neighbors defines the interaction volume, which is directly related to the resolution in the AFM and STM image. The same argument can be made if we transition to a non-ideal tip, where the side atoms of the tip are interacting with the sample. In order to achieve atomic resolution the interaction volume must be small, and second and higher nearest neighbor interactions must be minimized. In STM the current–distance curves are very steep due to the exponential decay of the tunneling current with tip–sample separation, nearest neighbor contributions are small, and atomic resolution therefore relatively easy to achieve. The interplay between attractive and repulsive, short and long range forces involved in dynamic force microscopy, which delivers atomic resolution on many insulating and conducting surfaces, is complex and they are discussed extensively by Giessibl [39].

2.1.3
Manipulation and Construction of Nanostructures with STM and AFM

STM and AFM are not only powerful tools for the analysis of surfaces and nanostructures, they can also be applied to manipulate ad-atoms and molecules, and control their motion and position on the surface. In imaging mode we usually strive to leave the system under investigation unperturbed, but it became evident early on that the probe techniques can be used to modulate surface structures and ad-atom motion in a controlled manner [40]. The most famous examples are probably the IBM logo assembled from Xe atoms on a Ni(110) surface, and the quantum corral [8, 23] made from Fe atoms on a Cu(111) surface, which for the first time visualized a standing electron wave within a circular potential well. The construction of these structures with atom level precision offers the ultimate nanostructure assembly. Atomic level assembly with scanning probes is a wonderful tool to study fundamental questions in nanoscience, but its drawbacks for applications are ultimately a very slow speed, and often the limited stability of the atom assembly against thermal excitation and motion. An atomic level computing scheme, where the presence and absence of Au ad-atoms in specific adsorption sites on the Si(111) (5×2) Au reconstruction denote a zero and one, was introduced by Bennewitz *et al.* [24] and an image of this system is shown in Figure 2.6. The advantages, for example, storage density, and disadvantages, for example, read and write speed, are discussed in detail and illustrate in this publication quite well the challenges of using true atomic-scale nanostructures.

The atomic level assembly of nanostructure with STM relies on the ability to control the motion of ad-atoms with the STM (or AFM) tip. The short-range forces between the tip and ad-atom drive lateral ad-atom motion, and their magnitude is controlled by the current and bias voltage, which determine the gap resistance and tip–surface distance. The surface diffusion of ad-atoms in the absence of an STM tip is controlled by the availability of adsorption sites, which are defined by the corrugation of the

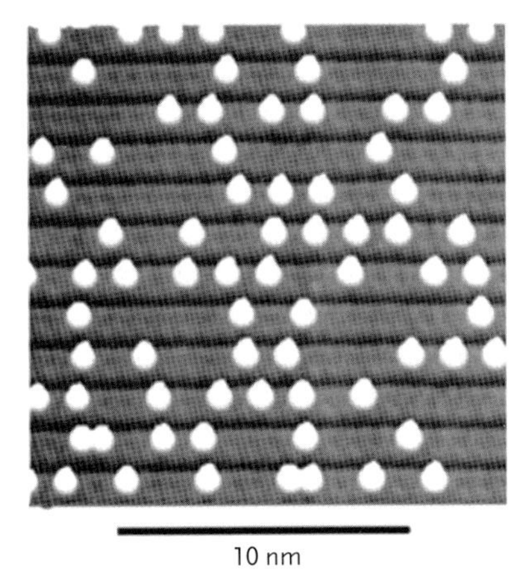

Figure 2.6 Atomic level computer [24], which is constructed by positioning Au atoms on a Si(111)5 × 2 reconstructed surface. This surface provides well-defined preferential bonding sites for Au ad-atoms, which can be moved by the STM tip thus defining a 1 (Au atom) or 0 (no Au atom). The write and read processes are discussed in detail in the reference. The figure is reprinted with permission from Nanotechnology [24] IOP Publishing Ltd. and R. Bennewitz (author).

surface itself. In order to move from one energetically favorable site to the next one, an ad-atom has to overcome an activation barrier, which is highly directional and specific for the respective surface structure. The movement of an ad-atom by the STM tip is consequently controlled by the subtle interplay between the tip–atom interaction and the energetic barriers defined by the surface structure.

Bartels *et al.* [40] explored this subtle interplay between tip–ad-atom and ad-atom–surface interactions in the motion of a wide range of adsorbates including CO, Pb, and Cu on a Cu(211) surface. This surface exhibits well-defined terraces, and grooves and thus provided an excellent testing ground to unravel the influence of the local energy barriers between preferential adsorption sites. In addition to ad-atom motion, which relies on the interaction forces, it is also possible to initiate chemical reactions through the local charge injection via the tunneling current. The removal of covalently bonded hydrogen from a Si(100) surface [41], for example, allows to selectively write conduction pathways of (2 × 1) reconstructed surface regions. In this nanoscale lithography process the hydrogen termination functions as the "resist" and the STM tip is the initiator to its removal at select positions. The injection of an electron leads to a weakening of the covalent Si–H bond, presumably through the excitation of vibrational modes, and is followed by desorption of the hydrogen atom. A last example, which shows STM as an active part in the design and manipulation of nanoscale structures, is the motion of the so-called nanocar [42]. The nanocar is a molecule, which was designed to demonstrate rolling motion on the nanoscale: the wheels are fullerene molecules attached to a rectangularly-shaped, relatively large molecule, which acts as the body of the car. The interaction with the STM tip was used to initiate movement across a surface, and indeed demonstrated rolling, car-like motion on a surface.

2.2 Photoelectron Spectroscopy and Electron Spectroscopy Techniques

Electron spectroscopies are among the most versatile analytical tools, and offer chemical sensitivity, which has remained mostly elusive in scanning probe methods. Electron spectroscopies [43, 44] include Auger and photoelectron spectroscopy (PES), which probe the occupied states, X-ray absorption spectroscopy and inverse photoelectron spectroscopy, which probe the unoccupied states, and spin-sensitive techniques, such as X-ray magnetic circular dichroism, and spin resolved photoemission. One of the disadvantages of electron spectroscopies has been a lack of spatial resolution, but substantial progress has been made in recent years, and high spatial resolution down to the nanoscale is now available for several techniques. High-resolution scanning Auger spectroscopy, and photoemission electron microscopy (PEEM) are just two of the techniques, which now allow analysis of composition, chemical bonding and electronic structure with a spatial resolution below 10 nm. In this chapter we focus on the fundamental aspects of electron spectroscopy and discuss in more detail photoelectron spectroscopy, which is one of the most versatile techniques.

Figure 2.7 illustrates schematically the emission of photoelectrons, and the Auger process, which is the response of the electronic system to the formation of a hole with a strong local Coulomb potential. In the photoelectric effect, which is the basis of PES, the absorption of a photon with energy $h\nu$ leads to the emission of an electron whose kinetic energy E_{kin} is determined by the energy balance

$$E_{kin} = h\nu - \phi - E_B \tag{2.2}$$

where ϕ is the material's work function and E_B is the binding energy of the electron, and given with respect to the Fermi energy of the system (in Figure 2.7 the binding energy of the $L_{2,3}$, the p-orbital, is used as an example). In the interpretation of photoelectron spectra the so-called three-step model has been applied successfully, and it subdivides the photoemission process into the following steps: firstly, the excitation of the photoelectron; secondly, transport of the photoelectron through the bulk; and thirdly, emission from the bulk into vacuum, where the electron energy distribution can conveniently be measured. The energy distribution $P(E, \omega)$ of the photoelectrons, [45] which are excited with photon energy $h\nu = \hbar\omega$ can be described by the following expression (assuming that the material in question is crystalline and $\vec{k}$ is therefore a quantum

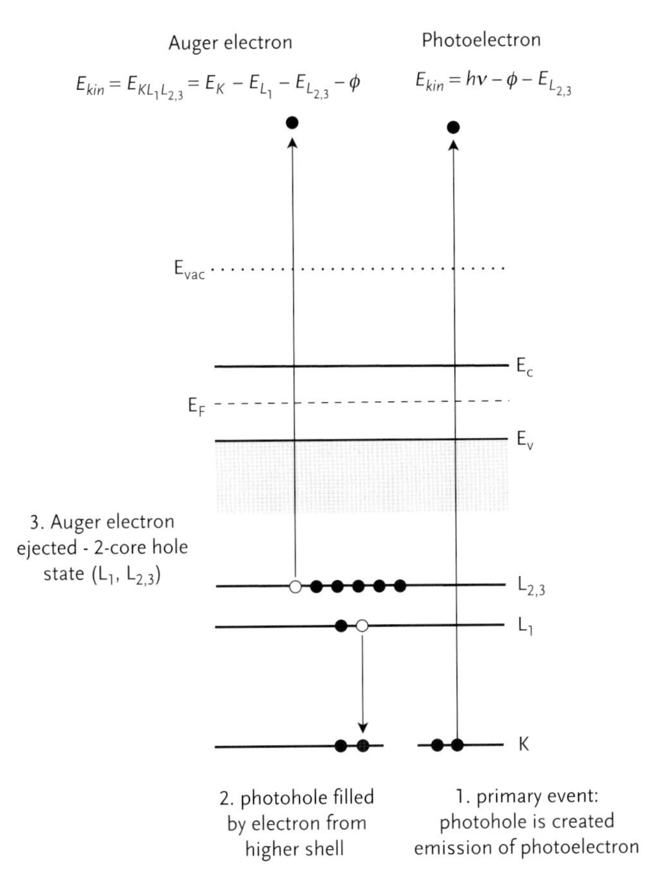

Figure 2.7 Schematic illustration of the photoelectron emission and Auger electron emission processes. The initial step is the excitation of a photoelectron from the K shell (in this example) and the creation of a photohole. This photohole can be filled by an electron dropping down from a higher lying energy level with simultaneous emission of a photon to conserve energy and momentum (X-ray emission), or the electronic system can relax by the emission of an Auger electron. The Auger process creates two core holes, and three energy levels are involved in the process, which is shown on the left hand side of the figure. E_c is the conduction band minimum and E_v denotes the position of the valence band maximum.

number to be conserved when considering excitations in the reduced zone scheme):

$$P(E,\omega) \propto \sum_{i,f} \int d^3k |M_{i,f}|^2 \cdot \delta[E_f(\vec{k}) - E_i(\vec{k}) - h\nu] \cdot \delta[E_i(\vec{k}) - E] \qquad (2.3)$$

where $|M_{i,f}|^2$ is the dipole transition matrix element, f and i are the final and initial states, which participate in the direct transition. The integral is only extended over the part of k-space containing the occupied states. The first delta function guarantees the energy and momentum conservation of the excitation process: E_f is the final and E_i the initial energy of the electron. The second delta function describes the measurement of the emitted electrons with a kinetic energy E. This expression is valid for direct excitations in crystalline materials. The k-conservation rule breaks down if the crystal symmetry is lost, k is no longer a relevant quantum number since it is defined with respect to the symmetry of the reciprocal lattice, and indirect transitions prevail. Amorphous solids like a-Si or a-C are examples of solids where crystal symmetry is lost. The resultant photocurrent is then given by the product of the initial and final state DOS weighted by an averaged transition matrix element.

The inelastic collisions, which electrons may experience on their way to the surface are neglected in this formalism, but they nonetheless contribute to the photoemission spectrum in the form of a secondary electron background. Discrete peaks in the secondary electron background can be observed due the excitation of collective oscillations of the valence band electrons (plasmon) and excitation of interband transitions whose positions correspond to maxima in the imaginary part of the complex dielectric function. PES can probe the core levels as well as the valence band, and is one of the few methods which can be used to measure the band structure of bulk and surface.

The excitation of a photoelectron creates a hole, and thus a strong local Coulomb potential, which triggers a rearrangement of the electronic charge. The relaxation of the photohole can be achieved by an Auger process, or the emission of a photon (X-ray emission spectroscopy, XES). The Auger process, which is shown in Figure 2.7, dominates for light atoms, while XES becomes more important in heavy atoms. In the Auger process, an electron from a higher energy level relaxes and fills the original hole state, creating a secondary hole in this higher energy level. This process is either accompanied by the emission of a photon to satisfy energy conservation (XES), or the ejection of a second electron (Auger electron). The kinetic energy of the electron is determined by the energy of the levels involved in the process, and is thus element specific.

The analysis of core level spectra in PES begins with the identification of core level peaks, and analysis of elemental composition [46]. The determination of element concentrations depends on relative excitation cross sections, which are element specific and tabulated, and the depth with respect to the surface in which the respective element is found. PES is highly surface sensitive, and its information depth is about three times the electron mean free path at the respective kinetic energy of the photoelectron. The electron mean free path is visualized in the so-called universal curve, which shows a minimum at an electron energy of about 40–50 eV. The low energy branch (<50 eV) is controlled by the electron–phonon scattering events, and the high-energy branch is determined by electron–electron interaction and its slope is less steep than that of the low energy side. However, an electron, which is excited at a larger depth is less likely to penetrate the surface and will thus contribute less to the photocurrent. The Lambert–Beer law describes the exponential dependence of the current (I) as a function of the ratio of the excitation depth (d) the mean free path ($\lambda_{electron}$) to: $I = I_0 \cdot e^{-d/\lambda_{electron}}$, where I_0 is the total number of excited photoelectrons prior to interaction with the bulk material.

The next step in an analysis is the determination of chemical shifts, where a variation in the local bonding environment can lead to a shift in the binding energy of the core level by up to several eV. As a general rule, bonding to an electronegative bonding partner will lead to a shift to higher binding energy (lower kinetic energy), although this can be superseded by complex responses of the electronic system to bonding. The energetic position of the core level is therefore used to probe the local bonding environment, and many of the chemical shifts are tabulated[1] and have been investigated in considerable detail. For the fitting of the core level peaks a Doniach–Sunic function [47] is often used, which combines a Lorentzian component related to the core hole lifetime, and a Gaussian component, which includes the resolution of the instrument. The Doniach–Sunic function [47] also includes an asymmetry parameter, which is required to correctly describe metallic systems, where the outgoing photoelectron excites electronic fluctuations at the Fermi energy, leading to a shallower peak slope on the high binding energy side as a consequence of energy losses of the photoelectron. The background is usually described with a Shirley background. The position of the core level peak is determined by the complex interplay between local bonding configuration and the interaction of the single atom with the solid. This discussion however, leads us into the details of the photoemission process and the relaxation of the core hole, and we refer the reader interested in this topic to the literature [43].

The chemical shifts in the ethylfluroaceteate molecule (CF_3–CO–CH_2–CH_3), or alkanes where one or more of the hydrogen atoms are substituted by fluorine, are the textbook example to illustrate the chemical shift of core levels. The carbon 1s (C1s) core level, which is positioned at a binding energy of 284.6 eV for alkane groups (e.g., –CH_3), is shifted by several eV to higher binding energy and the magnitude of the shift is proportional to the number of electronegative groups attached to the carbon atom. The presence of strongly electronegative groups leads to a charge depletion at the central carbon atom and thus an increase in the binding energy. This charge redistribution also initiates a chemical shift in the opposite direction to smaller binding energies in metal or semiconductor carbides. The chemical shift induced by electronegative (or -positive) bonding partners is an important aspect in the analysis of core level spectra, and affords invaluable information about material structure and bonding. However, it is not the only contribution to shifts in the binding energy of core levels, and we

[1] For example at http://srdata.nist.gov/xps/.

therefore discuss two examples, which offer additional insight into in the interpretation of binding energy shifts in core level spectra. The first example compares pure carbon materials and SiC, the second example discusses core level shifts in nanometer-sized Au clusters.

The first example, which is shown in Figure 2.8, is a discussion of core levels in carbonaceous materials, which can adopt a wide range of different structures, including amorphous materials, and the crystalline allotropes diamond and graphite. The different allotropes of carbon are discussed in detail in Chapter 6. Carbon is exceptional in many respects, and is particularly suitable to illustrate the impact of structure on the core level spectra. The differences in the shape and position of the C1s peaks for the different allotropes are substantial, and the spectra for diamond (purely sp^3 bonded crystalline), a-C (sp^2 bonded amorphous), ta-C (sp^3 bonded amorphous carbon matrix with sp^2 admixtures), and graphite (sp^2 bonded crystalline) are included in Figure 2.8. The variation in bonding between the

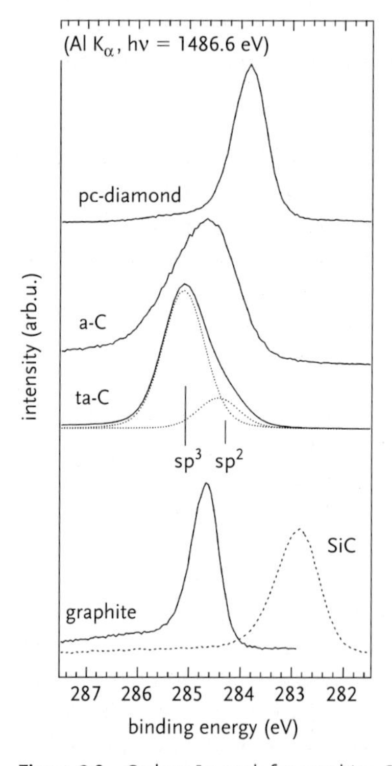

Figure 2.8 Carbon 1s peak for graphite, SiC, two amorphous phases (a-C, and ta-C) with different contributions from sp^2 and sp^3 hybridized carbon atoms, and diamond (polycrystalline diamond film). The modulation of peak widths and positions are discussed in the text. The photon energy of 1486.6 eV, which is used for excitation of the photoelectrons, is the aluminum K_α radiation emitted by a laboratory X-ray source.

sp^2 graphite-like and sp^3 diamond-like carbon atoms leads to a chemical shift, which is easily recognized in the spectra, despite the lack of a difference in electronegativity. The C1s peak of the amorphous materials is significantly broader than the C1s of the crystalline allotropes, which is attributed to small variations in bonding environment. The chemical shift between sp^2 and sp^3 carbon atoms can be used to unravel the relative contributions of sp^2 and sp^3 carbon atoms in the ta-C material.

The position of the C1s peak in diamond, however, does not conform to the assignments we just made using graphite, a-C and ta-C for the identification of chemical shifts. Diamond is a wide band gap semiconductor, and band bending is usually present due to the pinning of the Fermi level by surface states, and adsorbates. The Fermi level is therefore at a different position at the surface than it is in the bulk and the photoelectron will experience this potential difference, and the C1s binding energy (or more precisely: the kinetic energy of the photoelectron measured after it leaves the sample) is modified. The shifts of the diamond C1s peak, which are induced by surface band bending can be up to 1 eV, and the position of the diamond C1s therefore differs greatly between experiments. Carbon is indeed a material where the core level binding energies are exceptionally sensitive to structural variations, and chemical shifts: the electrons are covalently bonded, and not delocalized as is the case for metallic bonding, and the electronic system at a central carbon atom is therefore sensitive to its next nearest neighbor environment. At the other end of the spectrum are ionic solids, where the electrons are strongly localized and the local environment has little impact on the electronic system of the ion.

The second example is directly applicable to the analysis of nanostructures, and illustrates the shift in binding energy for the Au4f core level as a function of size for Au clusters with a diameter below 10 nm [50]. Figure 2.9 shows on the right hand side the Au4f core level for these clusters, and the two peaks are from the spin-orbit splitting of the core level: $Au4f_{7/2}$ (84 eV), and $Au4f_{5/2}$ (87.8 eV). The relative intensities of the two peaks reflects the occupancy of the energy levels. The binding energy of the core levels increases with decreasing cluster diameter, and the binding energy shift is inversely proportional to the cluster diameter. The origin of this shift lies in the dynamics of the photoelectron excitation process: it is usually assumed that the photohole is filled very rapidly (10^{-15} s) after the excitation of the photoelectron. However, this is only true if the system is sufficiently conducting. If we now place a metallic cluster on a poorly conducting substrate such as amorphous carbon, or SiO_2, the photohole will take a much longer time to be refilled and the charge becomes delocalized on the metallic cluster. The cluster can now be understood as a charged, spherical capacitor and the photoelectron leaving the cluster has to overcome an additional Coulomb energy, thus diminishing its kinetic energy, and increasing the binding energy of the core level peak. The Coulomb energy is inversely proportional to cluster diameter and capacitance, which is indicated in Figure 2.9.

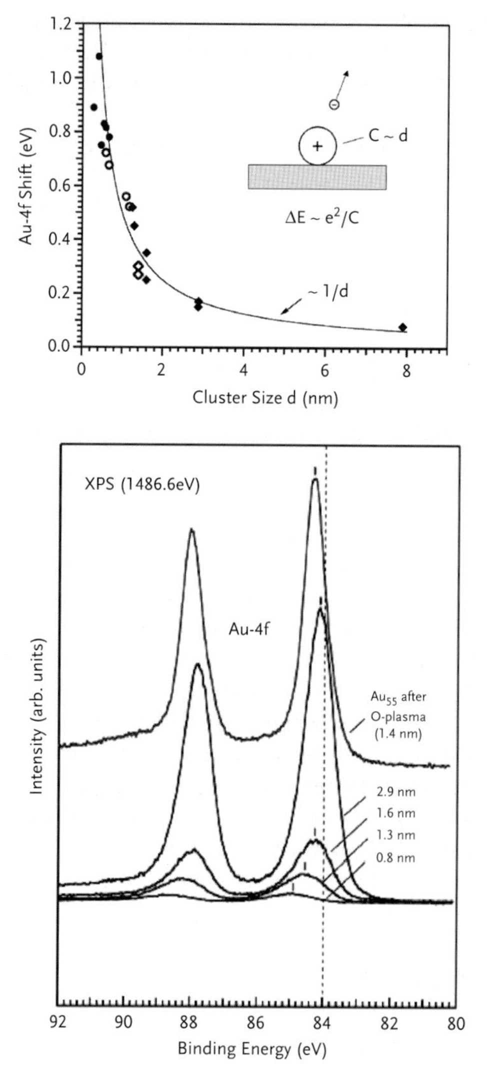

Figure 2.9 Au4f core level of Au clusters [50], which are synthesized with the micellar technique; the Au clusters are surrounded with a layer of micelles (organic molecules with hydrophobic and hydrophilic end groups), which stabilize the nanosize clusters. This is one of the few methods, which allows the size selected fabrication of sub-nanometer clusters such as Au_{55}, and it is described in more detail in Chapter 4. The spectra in the bottom panel show clearly the shift of the core level as a function of the cluster size: as the size is increased the Au4f peak moves towards the bulk value, which is indicated by the broken line. The top spectrum is measured after the removal of the ligand shell by means of an oxygen plasma. The Au clusters are immobilized on a SiO_2–Si surface. The graph in the top panel summarizes the core level shifts with respect to the bulk value (set to zero). The symbols refer to clusters prepared by different methods: full squares are the clusters which correspond to the dataset shown in the bottom panel, the open square is the Au cluster after removal of the micelle shell (corresponds to the core level observed after O-plasma shown in bottom panel), open circles and full circles are size-selected clusters on amorphous carbon, and SiO_2–Si, respectively. The core level shift is due to a charging of the cluster, which can be treated as a spherical condensor. The line represents the charging energy as a function of cluster diameter. Reprinted figure with permission from [50]. Copyright 2005 by the American Physical Society.

The Coulomb charging energy is also discussed in the context of the single electron transistor in Chapter 7.1. The binding energy shift in the core level can easily be measured in the XPS spectra and is an excellent measure of cluster size.

While photoelectron spectroscopy of the core level is certainly the most frequent use of the PES technique, we want to briefly describe here the spectroscopy of the valence band system, due to its importance in the measurement of band structures. It has played an important role in unraveling the electronic structure of graphene, and confirmed its peculiar band structure with a linear (conical) *E–k* relation in the vicinity of the Fermi energy at the six corners of the 2D Brillouin zone (see Chapter 6). The *E–k* dispersion curves meet at these six corners and form a so-called Dirac point, where the effective mass of the charge carrier approaches zero. Opening a band gap in the graphene band structure, and doping are among the critical features which must be realized in order to use graphene as an electronic material. The PES of the valence band has been instrumental in studying different pathways to achieve these goals, and they are described in more detail in Chapter 6. Another example in this context is the measurement of the one-dimensional band structure of horizontal nanowires discussed in Chapter 3.

In angle resolved ultraviolet photoelectron spectroscopy (ARUPS) the kinetic energy of the valence electrons and their corresponding wavevector are measured by probing different angles with respect to the surface normal. The photon energy is in the UV energy range (10–100 eV) in order to probe the dispersion of valence bands. The wave vector k is composed of a component parallel, $k_{||}$, and a component perpendicular, $k_{\perp}$, to the surface. In analogy to the transmission of a photon through a boundary between materials with different refractive indices, only the wave vector component $k_{||}$ is preserved when the electron is transmitted through the surface. $K_{||}$ is related to the electrons kinetic energy by

$$k_{||} = k \sin\theta = \left[\frac{2mE_{kin}}{\hbar^2}\right] \sin\theta \tag{2.4}$$

and θ is the angle with respect to the surface normal. Electrons are detected within a well-defined solid angle, and thus the spatial relation to the crystal structure of the solid is known. The kinetic energy of the electron can then be traced back to the respective wave vector k. From these data the band structure ($E(k)$) of surface and bulk states can be constructed.

2.3 Closing Remarks

This chapter has introduced several experimental techniques, which are important for the reader of this book, but it is only possible to scratch the surface: the number of techniques available to date is vast. New analytical techniques are developed rapidly and enable us to examine materials in new ways, with higher resolution and greater speed, thus deepening our comprehension of structure–property relationships. The capabilities of new techniques extend far beyond this short introduction, and we hope that at least some of the readers of this book will have the opportunities to explore these exciting possibilities in their own research.

3
Semiconductor Nanowires

Semiconductor nanowires are among the most versatile inorganic nanostructures, and can be fabricated from a wide range of materials. They are solid structures with a very high aspect ratio, and are thus distinct from hollow nanotubes, which are discussed in Chapter 6. Nanowires were among the first nanostructures, which had already been grown by the mid-twentieth century, although at that time they were usually classified as whiskers [51–57] and their diameters were still relatively large frequently reaching a few hundred nanometers. In the meantime nanowires have been developed into true nanoscale materials, and electronic devices have been built on single nanowires by spatially selective doping, and connectivity can be introduced through selective growth of branched structures. The design of nanowire circuits has progressed at a fast pace in recent years, and many exciting challenges remain before large-scale production of nanowire circuits can be realized. However, their use extends far beyond nanoscale electronics, and includes but is by no means limited to, highly selective sensors, efficient photon absorbers in solar cells, and optical applications [58–62]. These applications rest on the one-dimensional (1D) geometry and high aspect ratio of the wires, quantum confinement, and a large surface area, which is particularly beneficial for sensor applications.

Figure 3.1 gives an impression of the variability in semiconductor nanowire materials and please take a moment to look at the figures in this chapter to get a first impression of the diversity of nanowire structures. This figure should be understood only as a rough guide and cannot claim to represent the whole breadth of nanowire materials which have been synthesized to date, and only wires made of one or two atomic species are included. The semiconductor nanowires can be organized according to their composition, crystallographic orientation, morphology (e.g., branched, straight, . . .), composition along and across the wire (e.g., core shell structures, doping profiles), and dimensions (diameter, length). This scheme helps us to grasp the incredible variability within this class of inorganic nanostructures. The techniques used in the growth of nanowires have been expanded and adapted to the formation of a wide class of structures which are morphologically related to nanowires, such as nanobelts and nanorods.

Inorganic Nanostructures: Properties and Characterization, First Edition. Reinke, P.

Published 2012 by WILEY-VCH Verlag GmbH & Co. KGaA

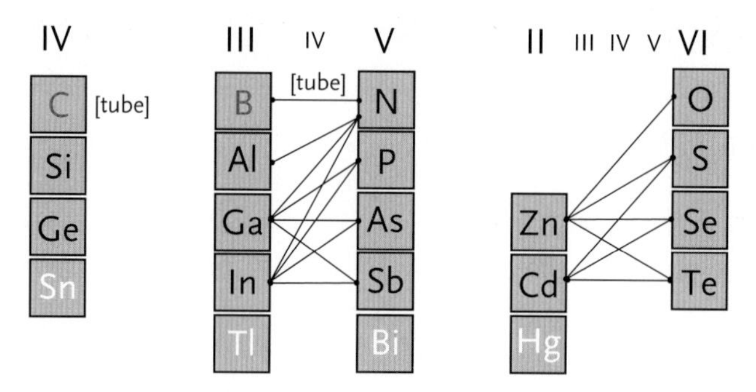

Figure 3.1 Sections of the periodic table to illustrate the composition of the most common semiconductor nanowires: group IV, group III–V, and group II–VI compound semiconductors. Carbon and BN adopt a tube geometry with a hollow core and are introduced in Chapter 6.

Semiconductor nanowires, which are the focus of this chapter, are free-standing and only attached to the substrate at one end of the wire, and we concentrate on group IV and III–V nanowires. By making this selection we can discuss the most critical aspects of semiconductor wire synthesis, properties, doping, and electronic structure without being distracted by the fascinating diversity of nanowire materials. The only other group of 1D nanostructures included in this chapter are horizontal nanowires, which "lie down" with the long axis parallel to the surface. These supported, horizontal nanowires include metallic and semiconducting materials and are often considerably smaller, both shorter and thinner, than the free-standing wires, and are strongly coupled to the underlying substrate. Even a few examples of truly monoatomic, supported wires have been reported.

3.1 Nanowire Growth

Probably the most important technique for the synthesis of nanowires is derived from the vapor–liquid–solid (VLS) growth of whiskers, which was demonstrated nearly half a century ago [53, 63]. The VLS method is highly versatile, relatively easy to implement and applicable to a wide range of material combinations of substrate and wire. Full comprehension of the thermodynamic and kinetic aspects which drive the VLS growth, is critical to achieving the desired control of nanowire properties, but many fundamental aspects of the growth process are still the subject of intense discussion in the literature. Other vapor-phase based growth methods include vapor–solid–solid (VSS) growth [64], and oxide-assisted growth [65–67], which does not require a metal catalyst. The solution-based methods for wire synthesis are firmly rooted in synthetic chemistry and will not be discussed here in detail, although some aspects of solution-based processing are discussed in the context of quantum dot and metal cluster synthesis (Chapters 4 and 5).

3.2 Vapor–Liquid–Solid and Vapor–Solid–Solid Growth

The VLS growth process [53, 55, 63, 68] is started by the deposition of a catalyst, which is a nanosize metal particle positioned on the substrate surface, and as we will discuss later, the size and position of the catalyst particles is critical in achieving control over the nanowire population. Figure 3.2 summarizes the VLS growth process for a Si nanowire from a Au catalyst [53]. The Au–Si phase diagram, which is the basis for understanding the progression of the nanowire growth, is included in the figure [150]. The deposition of catalyst particles is followed by their exposure to the gaseous precursor for nanowire growth. Generation of the gaseous precursor can be achieved by a wide range of techniques including thermal evaporation of a solid, introduction of a molecule by chemical vapor deposition (CVD) or a metal–organic complex by metal–organic vapor deposition (MOCVD), and laser ablation.

The Au–Si system is one of the most thoroughly studied catalyst–wire systems and as such an excellent example to discuss many aspects of VLS growth. The phase diagram of the Au–Si system shown in Figure 3.2 represents a typical binary phase diagram with a eutectic. The eutectic temperature T_e is at 636.15 K (363 °C) on the Au-rich side of the phase diagram and a liquid catalyst droplet can be sustained if the growth temperature is chosen to be at or above T_e. The reaction sequence begins with the pure, solid Au particle on the left hand side of the phase diagram. The precursor for the wire material which arrives from the gas phase, then begins to dissolve in the Au particle and a mixture of liquid Au precursor and solid Au is formed. Once the precursor concentration is sufficiently high, the mixture crosses the liquidus line and the Au precursor particle liquifies, and stays liquid until the precursor concentration is high enough to cross the second liquidus line, at which point the precursor itself solidifies and the nanowire begins to nucleate. Once the nanowire has nucleated it continues to grow and the precursor on the liquid catalyst/precursor particle is replenished from the gas phase. The liquid phase can sustain relatively high fluctuations in the concentration of the precursor and the growth is continuous.

In most cases the diameter of the catalyst droplet defines the wire diameter, and as such it is critical to control the position and size distribution of the catalyst in order to control the wire population. The observation of the growth process shows that the wire growth is usually initiated at the

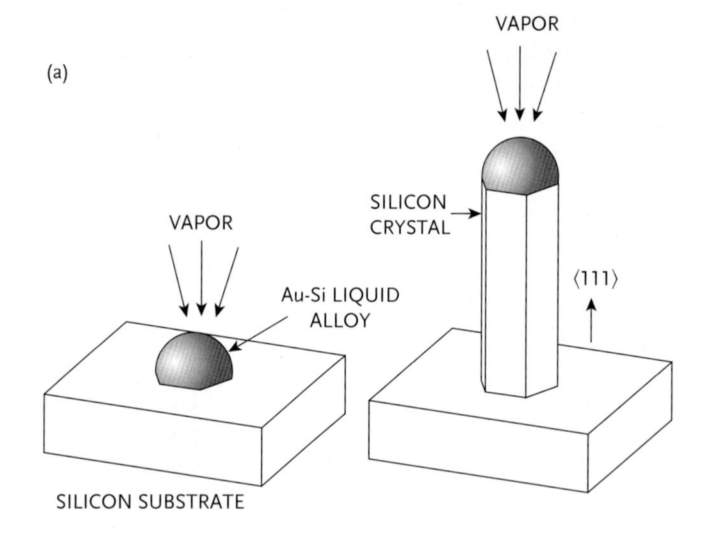

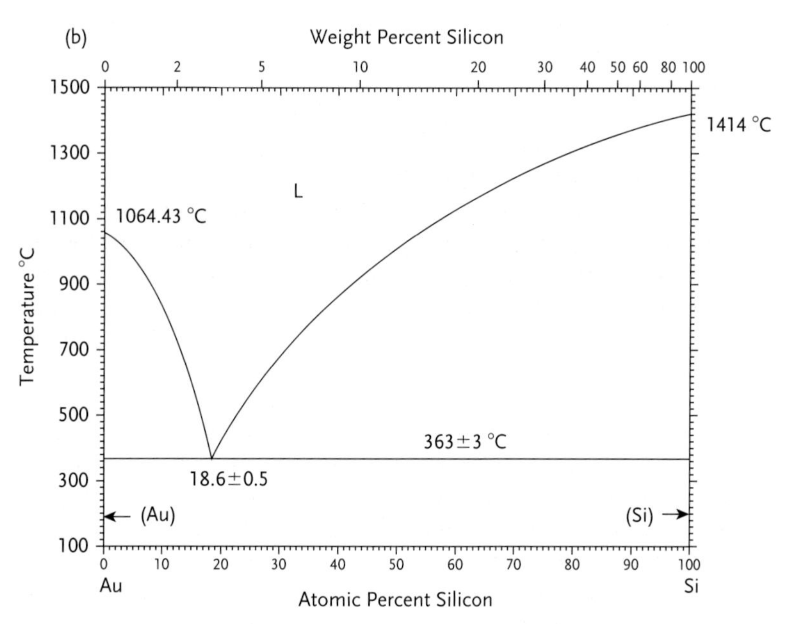

Figure 3.2 (a) Schematic illustration of the VLS process, which was proposed by Wagner and Ellis [53] and is still considered the best description for many catalyst-assisted nanowire growth processes; (b) Au–Si phase diagram [150] with the eutectic temperature T_e at 363 °C and a eutectic point at 18.6 at% Si. The schematics showing the VLS growth (a) is reprinted with permission from Applied Physics Letters [53]. Copyright 1964, American Institute of Physics. The phase diagram is from [150] and reprinted with permission of ASM International. All rights reserved. www.asminternational.org

substrate–catalyst interface, and then proceeds in a layer-by-layer mode at the wire–catalyst interface. The wire grows and the catalyst particle is moved along with the wire tip and nanowires are therefore frequently terminated by the catalyst droplet.

The phase diagram serves as a guide for the selection of a suitable catalyst and the initial choice of growth parameters, but the thermodynamic properties of nanoscale systems can differ substantially from the bulk properties represented in tabulated phase diagrams. Solubilities, enthalpies of mixing, the exact position of the phase boundaries, and other critical thermodynamic properties can change for small particles, and R.T. deHoff gives a concise treatment of these effects in his thermodynamics textbook [69]. Unfortunately the impact of these changes is often not known and most models treating the thermodynamics and kinetics of nanowire growth must resort to the use of well-known bulk phase diagrams. An example in this context is the extensively studied melting point depression of metal particles as a function of size [70]. It was shown early on that the melting point of Au particles decreases rapidly with particle size [70–73]. The reduction of the melting point is considerable and begins at a cluster diameter of about 10–20 nm and melting will occur if the surface free energy of the molten particle is less than the surface free energy of the solid particle. The relevant energy and entropy terms can be described in the framework of thermodynamics, after including additional terms, which account for curvature related changes in the free energies. Several studies of melting point depression illustrate the complexity of the process: (i) the particle shape/geometry [73] influences surface area and bonding, and thus the resultant surface energy–phase transitions as a function of particle size can therefore modify the melting process; (ii) supported clusters [71] have an interface with the substrate and the interfacial contribution to the surface free energy has to be taken into account; (iii) certain cluster sizes are exceptionally stable [72] (magic clusters, see Chapter 4) which favor the solid over the liquid phase and leads to discontinuity in melting point depression. The impact of these effects has to be assessed separately for each material and size regime, although as a rule of thumb, they will play a more critical role in the size regime below about 20 nm.

In many material systems the temperature where successful nanowire growth can be achieved is found to be below T_e. This observation lead to the proposition of an alternative mechanism of nanowire growth, the VSS mechanism [64]. The catalyst particle is solid and the wire begins to grow at the solid–solid interface between catalyst particle and substrate. The VSS mechanism has been confirmed for several material systems, and the growth of InAs wires is a particularly intriguing example [74]. The growth of the nanowires ceases once the In–Au particles become liquid: the growth can only proceed via a solid Au catalyst, and T_e of the In–Au system is the upper limiting temperature for wire synthesis.

Another example is the Al-catalyzed growth of Si nanowires, where the growth temperature for successful wire formation lies below T_e, which was seen as proof of VSS type growth. In the Al–Si system, T_e is at 577 °C and therefore considerable higher than in the Au–Si system. It was observed that Si nanowires can be grown at temperatures below T_e, and initially this was attributed to a transition from a VLS to VSS growth mechanism. However, a recent *in situ* transmission electron microscopy (TEM) study [75] refuted this idea and showed that the Al catalyst particles (diameter >50 nm) remain liquid far below T_e. The retention of the liquid state depends sensitively on the annealing and cooling protocol, and once the catalyst droplets solidify the wire growth is suppressed. The liquid state of the catalyst droplet below T_e is sustained due to supercooling. The Si wire growth with an Al catalyst therefore adheres to the VLS mechanism. It is interesting to note, that the VLS and the VSS growth methods are incredibly versatile and lead to nanowire synthesis for a very wide range of materials, but that each material combination presents a different set of fundamental reactions and processes, and thus challenges.

The description of the VLS process is often limited to a discussion of the phase transitions, an approach also taken in the previous paragraph, but this view fails to explain the preferential nucleation at the catalyst–substrate interface or the unidirectional growth of the wires. In order to arrive at a more adequate description of the VLS process, the interfaces, fluxes across the interface and surface, and local variation of chemical potentials must be examined in detail. A comprehensive analysis of the thermodynamic and kinetic aspects of nanowire growth in the VLS process is given in a recent review article by Wacaser *et al.* [76].

The catalyst–wire–substrate system can be subdivided into (i) the gas phase (supply phase); (ii) the supersaturated liquid catalyst droplet with a curved surface; (iii) the three-phase boundary at the catalyst droplet–wire interface; (iv) the two-phase interface between catalyst droplet and wire; and (v) the substrate and wire–substrate interface. The material transport (flux) across these interfaces is driven by the chemical potential gradient between gas phase, catalyst droplet and nanowire and guarantees that a supersaturation of precursor material is maintained in the catalyst droplet, which then drives the crystallization of the nanowire at the catalyst–wire interface. The overall growth conditions, including temperature, precursor flux from the gas phase, and relative sticking coefficients on all surfaces within the system, control the local supersaturation and thus determine the probability of successful formation of a nucleus at the substrate–catalyst interface. The conditions for successful VLS growth must be chosen such that they favor the nucleation at the catalyst–droplet interface over the nucleation on all other free surfaces. The subsequent growth of the wire at the catalyst–wire interface defines the growth surface, and the wire diameter is therefore controlled by the catalyst particle diameter, in agreement with experimental observations.

3.2.1
The Size and Position of the Catalyst Particle

The control of size and position of the catalyst particles are critical ingredients in the synthesis of nanowire arrays and functional, multiwire circuits since the catalyst particle defines the spatial distribution as well as the diameter of nanowires. The spatial and size distribution of the catalyst particles is essentially imprinted on the nanowires, although control over these two aspects of the catalyst distribution is rarely achieved simultaneously, and is not always necessary. The control of spatial distribution of catalyst particles can be achieved by lithography techniques, which require the addition of several, often complex, processing steps, and surface templating, where the surface topography is altered to provide preferential nucleation and attachment sites for the catalyst particles. It is even possible to trigger the growth of complex nanowire structures, such as branched and tree-like geometries [77] described in Section 3.3.3 if the catalyst particles are deposited on pre-existing nanowires.

The fabrication of size-selected catalyst particles is a formidable challenge, and methods which provide some of the narrowest particle size distributions are gas phase synthesis of metal particles by aerosol techniques [78], or use of cluster sources [79]. Gas phase synthesis relies on the aggregation of metal particles through the control of the collision rates in combination with rapid quenching of the metal particle growth through dilution of the gas or supersonic expansion. Metal nanoparticles and clusters, their structure and synthesis, are discussed in detail in Chapter 4. However, the majority of experiments use catalyst particles which are formed by metal deposition directly onto the substrate surface through physical vapor deposition (PVD) and related techniques from solid sources. This is much easier and cheaper to implement, and can be combined with many different techniques for nanowire growth. The metal particles form spontaneously because either the metal does not wet the substrate surface and thus forms droplets, or de-wetting is initiated during an annealing step. The width of the size distributions is consequently relatively large leading to an equally broad distribution of nanowire diameters.

Unfortunately the initial size distribution of the metal catalyst particles can be modified dramatically during heating to nanowire growth temperatures. Ostwald ripening and coarsening set in, and the extent to which these processes modify the size distribution depends on the specific material system. The Gibbs–Thompson equation [80] states that the chemical potential is inversely proportional to particle radius, and a chemical potential gradient therefore develops between particles of different size. The gradient in chemical potential drives ripening and leads to the dissolution of smaller and growth of larger particles.

In the early stages of nanowire growth the diffusion of catalyst material across the substrate surface dominates, but as soon as the nanowire begins to grow, the catalyst particle is modified if diffusion on the nanowire surface itself contributes. A recent experiment by Hannon *et al.* [81], summarized in Figure 3.3, illustrates this process very well: the smaller catalyst particles on top of nanowires shrink while the initially only slightly larger particles grow at their expense. This ripening process includes material transport along the wire surface as well as across the substrate. Shrinking of the catalyst particle, which is positioned on top of a nanowire leads to a reduction in the nanowire diameter, which then

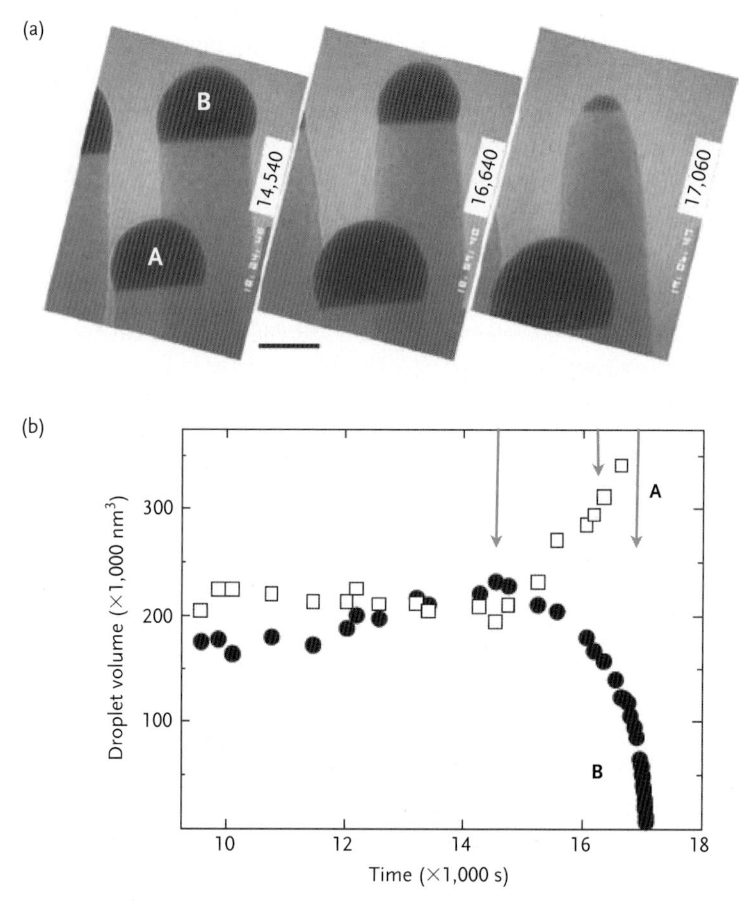

Figure 3.3 This figure illustrates the Ostwald ripening of catalyst particles, which are positioned on TOP of Si nanowires [81]. The TEM images in (a) were recorded inside a TEM chamber in real time. The catalyst (Au) diffuses along the nanowire surface and across the substrate and the ripening process progresses as the growth process continues. This process can modify the wire diameter quite substantially and has therefore a considerable impact on the homogeneity of the sample. (b) shows the catalyst particle volume as a function of time. The times corresponding to the images in (a) are marked by arrows. Reprinted by permission from Macmillan Publishers Ltd. Nature [81], copyright 2006.

becomes tapered at the top and ceases to grow when the catalyst particle has finally dissolved. Surface diffusion of the catalyst material along the wire surface can be suppressed in this case through the addition of oxygen to the gas phase. The modulation of the temperature-dependent surface transport can, on the other hand, be used to control the nanowire shape.

3.3 Nanowire Crystallography – Wire Structure

Two aspects of nanowire crystallography are critical to the application and integration of semiconductor nanowires: firstly, the crystallography of the wire itself, which is for the III–V and some II–VI wires driven by the interplay between the wurtzite (WZ) and zincblende (ZB) structure (Figure 3.4); and secondly, the crystallographic relation between substrate and semiconductor nanowire, which determines the growth direction of the nanowire. In bulk and thin films most III–V and II–VI semiconductors, with the exception of some nitrides, adopt the ZB structure while in wires the WZ structure often dominates. Nanowires in the III–V system, which grow in the $<111>$ direction are particularly prone to the formation of twin planes and stacking faults. The introduction of a stacking fault can lead to switching from the ZB to WZ structure or vice versa in a section of the wire, and the accumulation of this type of defect is detrimental to many aspects of wire performance within a device. Defects and switching of the structure

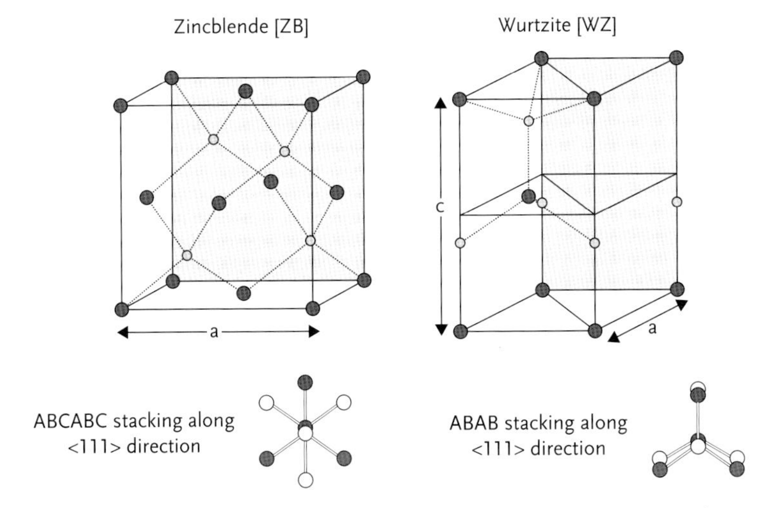

Figure 3.4 Crystal structure of zincblende and wurtzite (for details on crystal structure see [80]). The full circles represent the cation, the open circles the anion. The small diagrams depict the atom arrangement as it is seen when looking along the $<111>$ direction.

along the wire can lead to increased scattering of charge carriers, band gap modulation along the wire, and concomitant band offsets [82], and provide recombination centers, which diminish optical properties. Ideally, one would like to achieve full control of the wire crystallography and find conditions, which allow selective growth of WZ or ZB type wires or well-defined superlattice structures. Growth of the wires along the <100> direction greatly diminishes the accumulation of extended structural defect, but the <111> growth direction is usually preferred since it proceeds by nucleation and growth via the low energy (111) plane [11].

3.3.1
Competing Structures: Wurtzite and Zincblende

The WZ and ZB structures are shown in Figure 3.4. The ZB structure (sphalerite) is a cubic structure based on the face centered cubic (f.c.c.) lattice where every second tetrahedral interstice is occupied with an atom of the other kind. It is therefore a close relative to the diamond structure. The stacking along the body diagonal, the <111> direction, is an ABCABCA . . . stacking. The WZ structure is the hexagonal counterpart with a hexagonal closed packing and thus ABABA . . . stacking order, and again half of the tetrahedral sites are occupied. For the majority of III–V semiconductors the ZB structure is the more stable phase in the bulk, but the WZ structure dominates in nanowires below a material specific wire diameter; for some nitrides such as AlN, GaN the WZ structure is preferred in bulk materials. The cohesive energy difference for GaAs is about 24 meV per III–V atom pair in favor of the ZB bulk structure. If the cohesive energy difference becomes very small as for example in CdSe, or ZnS, it is possible to obtain both types of structures in the bulk phase. An extensive calculation of the energetic differences between the bulk WZ and ZB phases for binary octet compounds (IV–IV, III–V, II–VI), which are mostly ionically-bonded semiconductor materials, was performed by Yeh *et al.* [83].

The energy considerations that apply to the bulk cannot be transferred directly to thin films or nanowire structures. For example, while the ZB structure is more stable for III–V bulk material, nanowires grown with a <111> growth direction on (111)B substrates[1] very often grow in the WZ structure, or present a mixture of ZB and WZ (polytypism). The <111>B direction is the most frequently used substrate orientation for the growth of III–V nanowires, and the details of nanowire–substrate crystallography are discussed in Section 3.3.2. The WZ and ZB structures can easily be converted into each other by the introduction of stacking faults [84, 85]: in WZ the insertion of a plane (e.g., **C**), which is out of order with respect to the

[1] <111>B refers to an anion terminated surface (e.g., As-terminated surface in the case of GaAs), while <111>A is cation terminated.

stacking sequence switches the lattice to the ZB structure: ABABAB to AB*AB***C**BC with the segment around **C** being ZB stacking [*AB***C**]. Introduction of a single misplaced plane [**A**] leads for the ZB structure to the formation of a twin plane: ABCABC changes to ABC**A**CBA. Introduction of two will form a WZ segment: ABCABC changes to AB*C***A***CA* and *CA***C***A* is the WZ segment. The introduction of twin planes in the <111> direction is a well-known structural defect in f.c.c.-based structures and is indeed very frequently observed in nanowires. The challenges, which arise from these observations are deceptively simple: why is the WZ structure stabilized in nanowires, and how can the phase contributions be controlled? At present some aspects of these challenges are understood at least in an empirical manner, but more work is clearly required to achieve the desired structural control for the whole spectrum of nanowires.

The stability of the WZ structure in nanowires is generally attributed to a true nanoscale phenomenon: the increase in contributions from surface and edge energies to the total energy of the system. The surface energy differences between the ZB and WZ nanowire side facets can be estimated by counting under-coordinated surface atoms per area. Following this estimate the surface energy for the WZ nanowire is lower than that of a ZB wire with the same diameter. Akiyama *et al.* [86] followed this train of thought and calculated critical diameters below which the WZ phase dominates, taking into account facet and surface contributions of the complete wire. This model neglects reconstructions of sidewall facets, and local strain, which can be induced by the small curvature or facet size and might have sizeable contributions to the total surface energy.

A comparison with experimental values shows a considerable disparity: the theoretical values turn out to be far too small. For example, the experimental value [87] for the transition to a predominantly ZB wire for InAs is larger than 100 nm, while the theoretical value places this transition at a diameter of about 10 nm, a magnitude lower. The nanowire structure is therefore influenced by a much wider range of parameters, including the surface and interfacial energies, nucleation barriers and surface reconstructions. Many studies indicate that wire nucleation and growth at the catalyst–wire interface plays a decisive role in phase selection [55, 76, 84, 88–90]. A comprehensive, fundamental description of polytypism based on thermodynamic and kinetic considerations has not yet been achieved, however, the experimental control of twin plane density, and ZB (or WZ) segment size has advanced rapidly in recent years.

The small energy differences between surface and bulk energies of the respective WZ and ZB structures make the system highly susceptible to small fluctuations in concentration, density, reactant pressure, size, and supersaturation of the catalyst. While this makes it very difficult to establish a satisfactory thermodynamic and kinetic description of the system, it also opens the pathway to adjust wire crystallinity through careful selection of the growth parameters.

Polytypism and twin boundaries are not in all instances detrimental to wire performance, but can be used to fabricate superlattices whose functionality is based on the periodic modulation of properties within a nanostructure. The fabrication of superlattices has been used extensively in thin film materials since they offer the possibility of manipulating electron and phononic band structures, and have even been explored for the modulation of thermal transport. The relative phase contributions and twin plane densities are quite sensitive to the experimental parameters, which opens a pathway to control and modify wire crystallinity [84, 88, 91] along the growth axis within a single nanowire. For example, GaAs wires (diameter around 50 nm) grown at $T = 350\,^{\circ}\mathrm{C}$ have a ZB structure; the density of twin planes increases rapidly for increasing temperature and a complete switch to the WZ dominated wires occurs around $550\,^{\circ}\mathrm{C}$. The temperature dependence of phase preference was explained with differences in the energy barrier to the formation of a WZ or ZB stacking sequence at the catalyst–wire interface.

Caroff *et al.* [84] demonstrated recently two different type of superlattices, which are integrated in InAs wires: a periodic lattice of twin boundaries in ZB wires, and a superlattice of alternating ZB and WZ regions which is created by controlled introduction of stacking faults. An image of an InAs wire, which contains a twin plane superlattice is shown in Figure 3.5

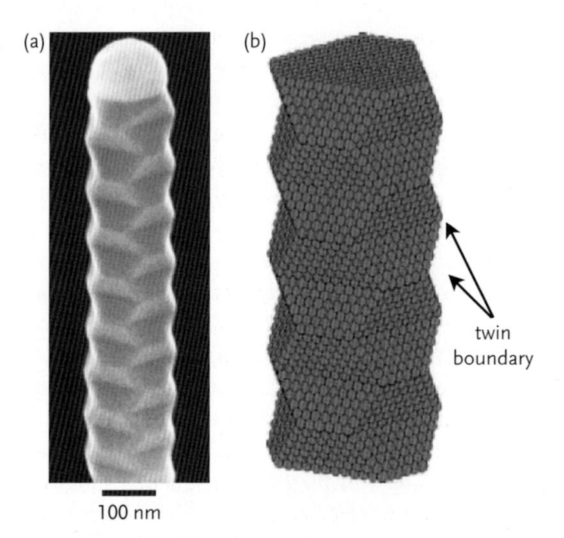

Figure 3.5 Superlattice of coherent twin planes on an InAs nanowire [84]. (a) is an SEM image of a wire and clearly shows the rotation of the facets and twin boundaries; (b) is a model of the twin superlattice and illustrates the arrangement of atoms within the wire. The twin planes and their impact on the nanowire habit can easily be identified. The side facets are in this case {111} oriented microfacets. The same article also shows the switch between ZB and WZ structures on a single wire. Reprinted by permission from Nature Nanotechnology [84] Macmillan Publishers Ltd., copyright 2008.

together with a model of the wire structure. The periodicity of the twin plane superlattice depends strongly on temperature and wire diameter, and can only be achieved for pure ZB wires, which form for diameters exceeding about 90 nm. Algra *et al.* [91] on the other hand triggered twin plane formation by adjusting the Zn concentration in the gas phase, which was introduced to act as an electronic dopant in InP wires. The ability to modulate on-wire crystallography offers an intriguing and highly versatile, although technically often difficult, approach to the fabrication of device structures, which are located on a single wire.

3.3.2 Nanowire Crystallography: Connecting to the Substrate

The preferred growth direction of most compound semiconductor nanowires is along the <111> direction of the ZB lattice, which is parallel to the <0001> direction in the hexagonal WZ lattice. The strong preference for the <111> growth direction is due to the nucleation of the low surface energy {111} facet at the substrate–catalyst interface. The <111> growth direction, however, is not necessarily the most desirable growth direction, since these wires often present a high density of stacking faults as described in Section 3.3.1, which can be detrimental to performance. Growth in the <100> direction is not only less prone to the incorporation of defects, but also more easily integrated in semiconductor processing.

The wires usually have an epitaxial[2] relation to the substrate, as the crystallographic directions from the substrate are transferred to the nanowire. The non-vertical growth of a nanowire ensues if the preferred growth direction is not normal with respect to the equivalent orientation at the substrate. The preferred growth direction can sometimes be influenced by the deposition parameters, or can switch during growth, due to subtle changes in surface energies or in the catalyst–wire interaction. If the substrate itself is polycrystalline or amorphous, for example, in an oxidized Si surface, the wire direction will be somewhat random and the strong directionality of growth as it is observed in epitaxial systems is lost.

A summary of experimentally observed growth directions for GaAs wires on four differently oriented GaAs substrates [92] is given in Figure 3.6. Many of these observations can be transferred directly to a much wider range of nanowire materials, including most III–V and IV nanowires. The III–V nanowires will grow vertically, parallel to the surface normal, if a (111)B oriented III–V substrate is used in the process. On a (001)

[2] Epitaxial growth is defined as a growth process where the crystal structure of the overlayer (in our case the nanowire) has a defined orientational relationship to the lattice of the substrate. However, this does not imply that the overlayer and substrate lattices have to be identical, it only means that they must possess a well-defined relation to each other.

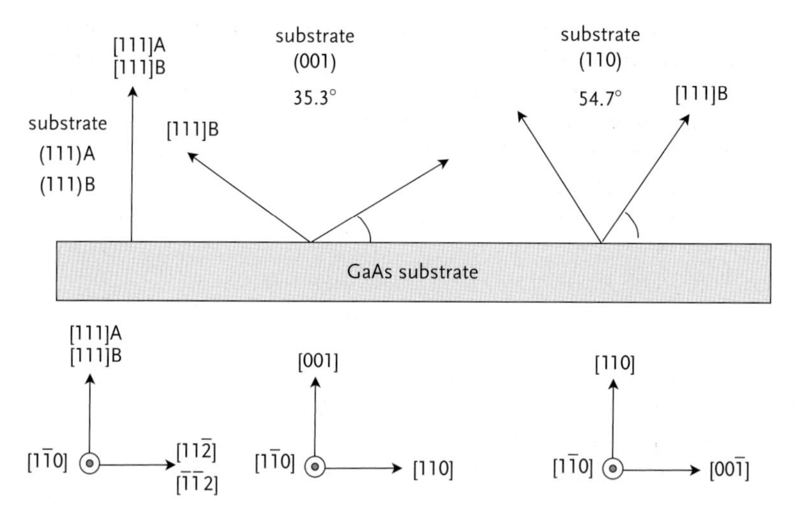

Figure 3.6 Summary of nanowire orientations with respect to the substrate. In all the wires the $<111>$ is the preferred growth direction and the epitaxial relation between the low energy (111) plane of the nanowire and the (111) plane of the substrate determines the relative angle between nanowire and substrate surface. These data are supported by a large number of investigations cited in this chapter. The figure summarizes many of the results described in [92].

oriented substrate the nanowire will then grow in the $<111>$ direction and acquire an angle of 35.3° with respect to the substrate surface. The family of $<111>$ lines yields four equivalent growth directions, which are separated by 90° in the horizontal plane. The growth on a (110) surface leads to a wire with an angle of 54.7° with respect to the surface plane. These crystallographic relations between the substrate and nanowire are valid if the preferred $<111>$ growth direction is sustained throughout the growth process, but exceptions to this general rule have been reported, although the underlying reason for a change in direction is often unknown. Switching of the preferred growth direction has, for example, been observed for InP on InP (001) substrates [93]: the wire first grows parallel to the surface, which corresponds to the $[1\bar{1}0]$ direction, and then switches to growth in the $[1\bar{1}1]$ direction which again has an angle of 35.3° with respect to the substrate surface. A similar observation was made by Mikkelsen *et al.* [94] for GaAs wires on GaAs (001) substrates, where the wire structure was observed for all sections by cross-sectional scanning tunneling microscopy (STM). This analysis even revealed a relatively complex faceting of the initial wire segment, which grows much like the InP wires in the $[1\bar{1}0]$ direction parallel to the substrate, and then switches growth to the $[1\bar{1}1]$ direction.

The growth of silicon wires on Si(100) and Si(111) substrates, and Ge wires on Si and Ge substrates [95, 96] mostly follows the crystallographic relations between nanowire and substrate laid out for III–V compound semiconductors. Si and Ge both crystallize in the diamond cubic structure, the nanowires grow epitaxially, and the <111> direction is again the preferred direction of growth. Consequently a Si wire will grow vertically on a Si(111) substrate and with an angle of 35.3° to the substrate surface on a Si(100) substrate; the same crystallographic relation holds true for Ge wires. However, while Si substrates are the most important material in electronic devices, the Si nanowires lack the powerful optical properties of the direct band gap compound semiconductor nanowires.

The next challenge is consequently to grow compound semiconductor nanowires with a well-defined orientation on Si substrates to enable combination of the optical properties of the nanowires with electronic circuitry based on Si [97]. The growth of III–V nanowires such as InP, GaP or GaAs on Si was initially thought to be unattainable because of their relatively large lattice mismatch and sizeable differences in thermal expansion coefficient. In contrast to the growth of multilayer thin film structures, the large lattice and thermal expansion mismatches turned out not to be a great obstacle to nanowire growth: the small diameter of the nanowires means that the elastic energy due to lattice mismatch can be relieved at the side facets thus minimizing the build-up of strain in the nanowire. However, in order to achieve heteroepitaxial growth of nanowire on the substrate it is necessary to remove the native oxide layer, which is experimentally challenging especially for Si due to the relatively high temperatures required for desorption. The epitaxial growth of several compound semiconductors has been demonstrated, including InP on Si(100) [98] and Ge(100) [97] despite a lattice mismatch of 8.1% with Si(100).

The growth of semiconductor nanowires on a semiconductor substrate of a similar nature still dominates the current research but a recent example illustrates the feasibility in using a variety of substrates [99] if the symmetry elements of the crystallographic structure of the substrate and nanowire are carefully matched. The $\gamma LiAlO_2$ (100) surface matches the $[1\bar{1}0]$ orientation of the GaN WZ structure in relative lattice parameter and symmetry; in a similar manner the MgO (111) substrate surface matches the [001] orientation of GaN WZ. Interface matching is used to ascertain the feasibility of an epitaxial relation between substrate and nanowire, and drives the GaN nanowire growth selectively in the orthogonal directions with respect to the substrate surface.

3.3.3 Complex Nanowires: Branching, Co-axial and Axial Nanowires

The vision which stands behind much of the research in semiconductor nanowires, is to construct entire nanowire circuits [57]. These circuits possess the full functionality of electronic circuits and will use nanowires, where the function is embedded within the nanowire's structure and composition [51]. The advantages of such circuits are their small and compact set-up, their versatility, and the built-in functionality on single wires, which can be defined during growth thus eliminating many complex and costly lithography steps. A new on-wire functional circuit element can be tested by modulation of the nanowire growth process, rather than requiring a complete redesign of device units. However, in order to achieve a nanowire circuit it is necessary to grow complex nanowire structures, which include branched structures, co-axial wires, which are also sometimes called radial wires, where composition changes in the x–y plane making core-shell structures, and axial wires, where composition is modulated along the long axis of the wire. It is possible to created on-wire quantum dots by axial, compositional modulation, and an example is discussed in Chapter 7 (quantum dots are discussed in detail in Chapter 5).

The axial variation of composition and structure can be achieved by rapid modification of the growth parameters and precursor composition in order to obtain a sharp interface. The requisite experimental techniques and knowledge have already been developed for the growth of planar thin film heterostructures (quantum wells) and superlattices. For axial modulation of the on-wire composition it is necessary to remain in the VLS mode, where the wire growth occurs via the catalyst–wire interface [100]. However, a core-shell wire with radial variation in composition can only be achieved if growth on the surface of the core-wire dominates over growth at the catalyst–wire interface. Any modification of the reactivity of the core-wire surface such as processing temperature, composition of the precursor gas, or oxidation can potentially be used to switch from catalyst growth mode to surface growth mode [101]. The addition of a shell is used to fabricate heterostructures with well-defined band offsets for electron and hole confinement or separation, the reduction of trap states and recombination centers at a wire surface, or simply to protect a wire against oxidation.

The branched structures can be achieved by triggering a secondary nucleation event, either by a second seeding process of catalyst particles on mature nanowires [77], or by changing the composition of reactant solutions if the growth is done by colloidal wet chemical methods [102]. The tree-like branched structures made of GaP nanowires, which were produced by using a second seeding process of catalyst particles, are shown in Figure 3.7. The TEM micrograph shows the seamless transition and homoepitaxial interface between the "tree trunk" and the "tree branches". The catalyst particle has been transported from the seeding position on the "tree trunk" to the tip of the "branch" during growth.

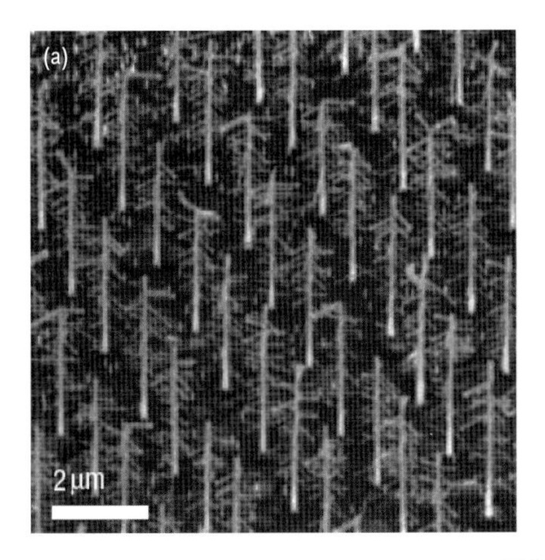

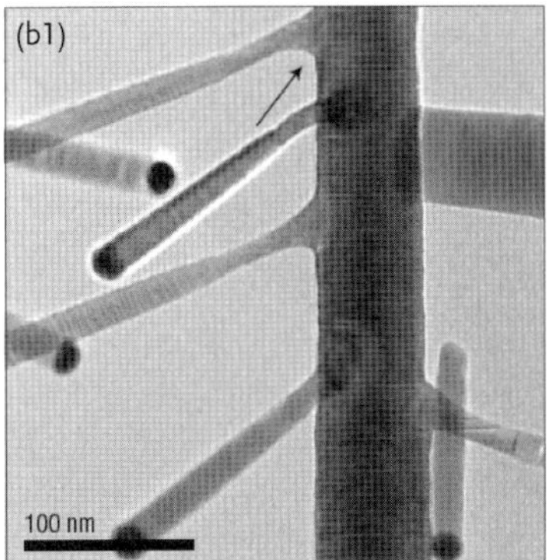

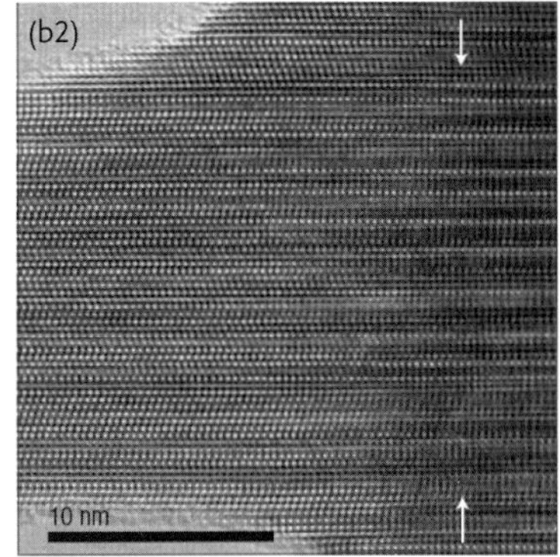

Figure 3.7 Example of a complex branched nanowire structure [77]. (a) SEM image of a "forest". (b1) and (b2) are TEM images, and the white arrows in (b2) indicate the seam between tree trunk and branch. The tree shape and formation of well-defined branches is achieved by deposition of a second round of catalyst particles on the original, vertical GaP nanowires. The direction of the branches is determined by the lattice matching between the tree trunk and the branch: the TEM image (b2) of the connection shows the continuity of the lattice and confirms a homoepitaxial growth of the branches. The catalyst is positioned at the end of the branches. Reprinted by permission from Nature Materials [77] Macmillan Publishers Ltd., copyright 2004.

3.4 Horizontal Nanowires

At the beginning of this chapter we introduced the distinction between vertical and horizontal nanowires. In the vertical wires a facet of the short dimension of the wires is attached to the substrate; in the horizontal wires, the long wire axis is parallel to the substrate and thus an intimate contact is formed. This 1D system can only be defined as an independent electronic unit if there is a potential barrier between wire and substrate, which is guaranteed if a metallic wire is placed on a semiconducting substrate, and the metallic states are positioned within the semiconductor band gap. In general, the driving force for horizontal wire formation is an inherent anisotropy in the interaction between substrate and deposit/ad-atoms. Anisotropy is therefore highly specific to the material combination, and can be due to an anisotropy in bonding sites, strain, or diffusion coefficients for ad-atoms.

Some of the best-known examples for horizontal nanowires are silicides, which form by a solid-state reaction between the Si substrate and the metallic silicide wire [103–110]. Their dimensions are typically between several nanometers in width and height, and up to a few micrometers in length. For some material combinations it is even possible to make monoatomic wires, which present the fundamental size limit for a 1D structure [111–122]. Most of the silicide wires are relatively good conductors and present only a small Schottky barrier to the Si substrate, and are therefore of great interest as nanoscale electrical leads. The fabrication of a true nanoscale metallic connector is one of the bottlenecks in the realization of nanoscale circuits, and carbon nanotubes, horizontal silicide wires, and some of the vertical wires discussed in the previous paragraph, are candidates for this fundamental building block. The monoatomic horizontal wires are, on the other hand, excellent systems to test fundamental concepts of electron transport and magnetism [112, 123–125].

3.4.1
Synthesis of Horizontal Wires

The synthesis of horizontal wires can be achieved in many different ways and is highly specific to the material combination. Several examples will be discussed in this chapter to illustrate the critical issues in the growth of wires, chains and other closely related 1D structures.[3] Many of these examples are 1D structures on the Si(100), Si(111) or a high-index Si surface, due to the incredible wealth of experimental work which has been performed for Si. The reconstructed Si(100)-(2×1) surface with its highly anisotropic dimer reconstruction, is an excellent template for the formation of many different kinds of horizontal wires, and Figure 3.8 shows sections of the periodic table to illustrate which elements form silicides and monoatomic wires on the Si(100)-(2×1) surface.

[3]The nomenclature of 1D systems on a surface is not consistent throughout the literature. Often a 1D row of atoms is called a chain, while the label "wire" is frequently used for a metallic, delocalized system. The 1D Bi structures were named Bi nanolines to stress their extraordinary length. We use the label "wire" for all extended 1D structures, unless a different name was explicitly chosen in the literature.

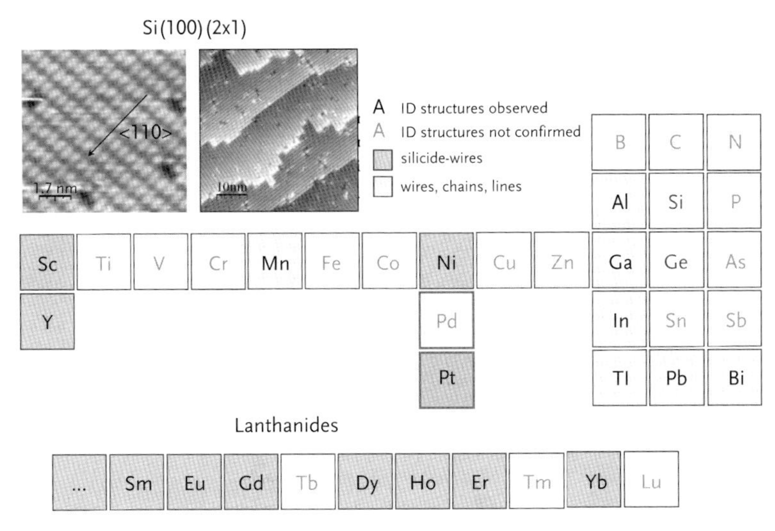

Figure 3.8 Summary of elements which form wires and chain structures on the Si(100)-(2 × 1) surface. An STM image of the surface is shown on the upper left corner and illustrates the dimer-row structure and asymmetry of the reconstruction. Elements, which form 1D monoatomic chains/wires are marked with a light gray background; elements which form silicide wires are marked in dark gray. The element designation is written in gray for those elements where it is currently not known whether they can form wires and chain structures. However, the literature on 1D structures in conjunction with the Si(100)-(2 × 1) surface is vast and this overview should therefore only be understood as a rough guide. Images courtesy of H. Liu, C.A. Nolph, and K.R. Simov.

The main pathways to the fabrication of 1D nanostructures, which are applied in the synthesis and design of a diverse set of 1D nanostructures, are (i) strain driven, highly anisotropic epitaxy as it is observed for the silicide wires [103–106, 108, 109, 126, 127]; (ii) templated self-assembly, which is driven by the substrate surface structure and often controlled by a specific reconstruction [114, 115, 119–121]; and (iii) step-edge decoration where the ad-atoms accumulate at step edges due to the enhanced step-edge reactivity [112, 117, 118].

The silicide wire formation is driven by a highly anisotropic lattice mismatch between the Si(100) surface and hexagonal silicide AlB_2 structure. This is illustrated in Figure 3.9 where hexagonal and tetragonal structures of $ErSi_2$ are shown [109, 110]. $ErSi_2$ is a typical example for a silicide structure, and the anisotropic mismatch is readily apparent by comparison with the geometry of the Si(100) surface structure. The anisotropy is greatly diminished for the tetragonal phase, which consequently grows as two-dimensional (2D) islands rather than 1D wires.

When the $(1\bar{1}00)$ plane of the silicide lattice is positioned parallel to the Si(001) surface, the [110] direction of the Si lattice is aligned with the [0001]

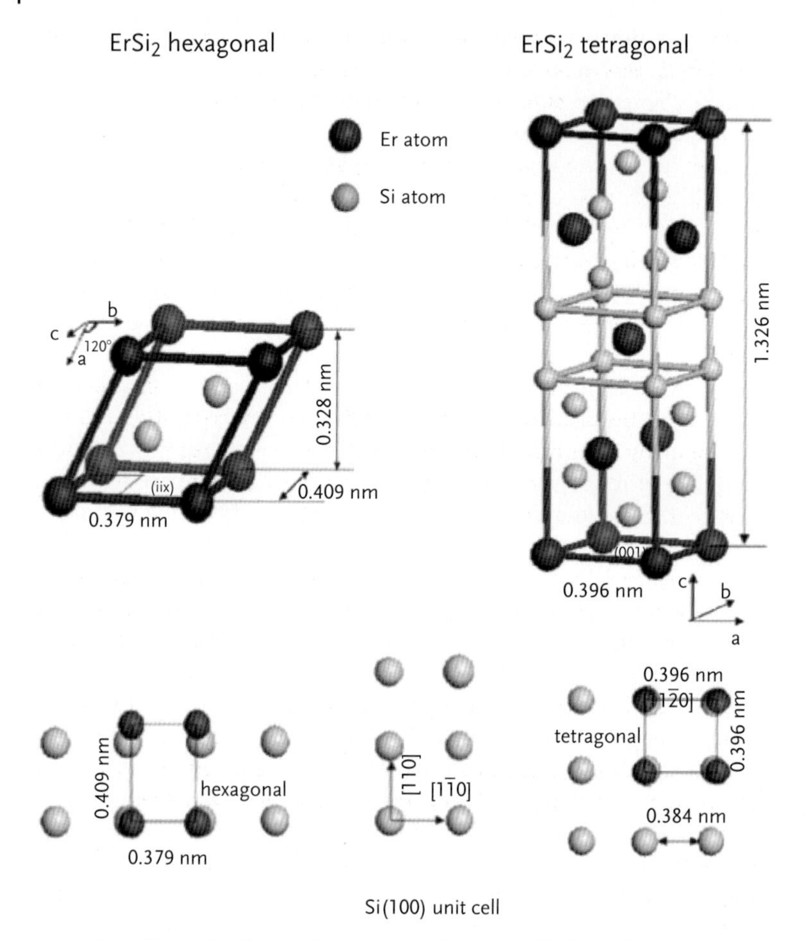

Figure 3.9 Illustration from reference [110] of the unit cells for the hexagonal and tetragonal AlB_2 structure of $ErSi_2$. The anisotropic lattice mismatch of the silicide unit cell with the Si(100) surface is the driving force for the growth of silicide wires in the hexagonal structure. This aspect is illustrated in the bottom row of the figure: in the center the unit cell of the Si(100) surface is depicted, and on the left and right the silicide unit cells of the hexagonal and tetragonal structures are projected onto the Si surface, respectively. The anisotropic lattice mismatch for the hexagonal phase is readily apparent. Reprinted with permission from Journal of Applied Physics [110] Copyright 2006, American Institute of Physics.

direction of the silicide. The [0001] direction (**c**(hexagonal phase) = 0.409 nm) of the silicide has a large lattice mismatch of several percent, while the $[11\bar{2}0]$ direction (**a**(hexagonal phase) = 0.379 nm) has only a small mismatch of at most 2%. The large anisotropy in the lattice mismatch breaks the symmetry of the system and leads to the highly anisotropic growth in the form of wires along well-defined direction of the Si(100) substrate. The length of the wires is limited by the probability of hitting another wire growing in the perpendicular direction, or if the growing wire front encounters a very high multilayer step edge, which forms due to step bunching. These step bunches are created during the growth of the silicide wires themselves: the weakly bound Si atoms from the step edges are consumed in the wire growth, which consequently leads to step bunching due to the anisotropy in step-edge modulation. The wire growth becomes self-limiting. The transition from the wire growth mode to an island growth, which is described in detail for $ErSi_2$ in [103, 109, 110], is due to a structural phase transition in the silicide from a hexagonal to a tetragonal lattice, and the large anisotropic lattice mismatch and associated strain is removed.

The growth of the silicide wires can be explained in a straightforward manner from the anistropy in lattice strain, but understanding the formation of many other types horizontal nanowires has to take into account a wider range of interactions. The surface itself provides a template, which can favor the formation of 1D structures due to specific bonding configurations, or a highly anisotropic diffusion of ad-atoms. While the silicide wires are relatively large, most horizontal wires are at most a few atoms wide, a few ten nanometers long, and approach the fundamental limit of monoatomic wire structures. The Bi-nanolines (Figure 3.10) are a notable exception and can grow to lengths of several hundred nanometers transecting several step edges on the Si(100)-surface [116, 128]. An STM image of Bi-nanolines on Si(100) is shown in Figure 3.10 together with the most recent model of its bonding geometry. The modification of the underlying Si substrate is quite substantial for the Bi-nanolines, which become partially embedded and are fully integrated in the surface reconstruction: the interface to the substrate is an integral part of the 1D wire.

The Si(100)-(2×1) reconstruction is an excellent template for the guided self-assembly of nanowires for a wide range of elements [118, 119, 121, 129–132] as shown in Figure 3.8. These studies covered mostly group III and IV elements, and computational work has given insight into the mechanisms of wire formation. The formation of many monoatomic wire structures on a free terrace can be described by the so-called anisotropic accommodation model. In this model the atoms stick preferentially to the wire end but are reflected from the side of the wire. The preference for attachment to the wire end can be a consequence of the local stress distribution, the stability of bonds within the wire, and bonding configuration with the substrate (e.g., the surface polymerization of Al wires). Monte-Carlo simulations of wire growth based on the anisotropic accommodation model were performed for Ga and In on Si(100)-(2×1) and are given in Refs. [121, 130–132]. Relatively short monoatomic wires (or chains) composed of magnetically active transition metals (Fe, Mn) and Au have also been produced by tip-assisted assembly in an STM, which allowed to study magnetic and electronic properties of well-defined 1D structures [133–135]. However, this single-atom approach has many technical challenges, and the guided self-assembly

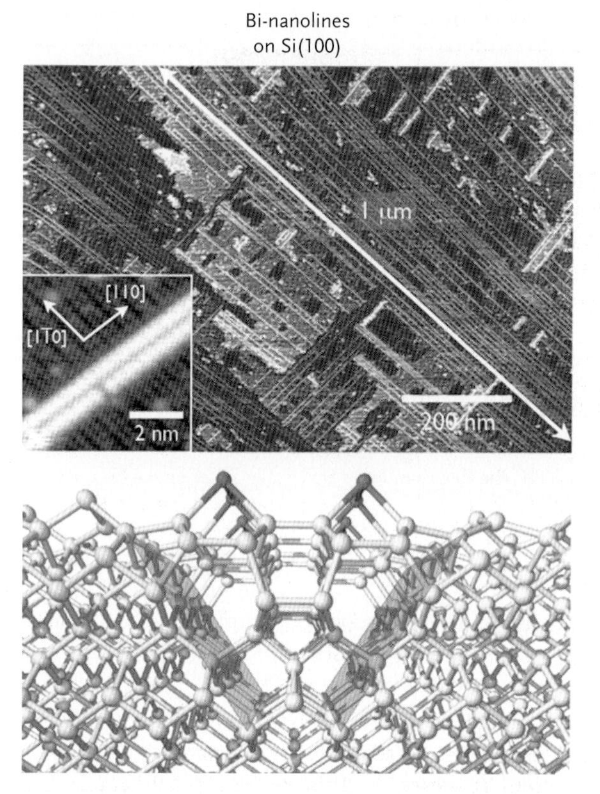

Figure 3.10 Top row: STM image of Bi-nanolines on Si(100); [116] bottom row: structural model. The structural model represent the current best knowledge of the wire structure, but due to challenges in determining atomic level structures it can be expected that it will have to be modified and adapted in the future. The Bi nanolines are exceptional in their stability, and can grow to lengths of several micrometers. The structural model for Bi bonding to the Si surface is complex, and the Si surface rearranges around the Bi-nanoline; the Bi atoms are marked in dark gray and sit on top of a five-membered protruding ring. With kind permission from Springer – Science and Business Media J. Mater. Sci. [116] copyright 2006, and kind permission from the author J.H.G. Owen.

offers a much more practical and scalable route to creating large-area nanowire assemblies.

One of the most versatile methods used to fabricate ultra-small wires is the use of a template with regularly spaced step edges, which can be produced on high-index surfaces. The high-index surfaces often fragment into microfacets, which are composed of a series of low-index terraces. The terraces are separated by step edges, and the terrace widths are defined by the crystal structure of the material. The terrace arrangement therefore depends on the selection of the high-index surface, and can be designed purely from crystallographic considerations based on tilt angle, tilt azimuth, and tilt zone, whereas the tilt zone is the low-index plane around

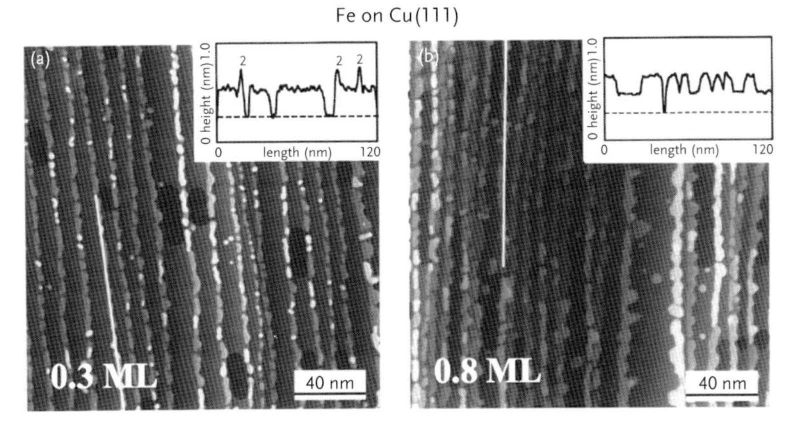

Figure 3.11 STM image showing step-edge decoration with sub-monolayer amounts of Fe on a Cu(111) surface [112]. The Fe accumulates at the step edges on the Cu surface, and grows into the terraces for increasing coverage. Reprinted figure with permission from Physical Review B [112]. Copyright 1997 by the American Physical Society.

which the rotation/tilt is performed. The expected terraces as a function of tilt and rotation can be read from a stereographic triangle such as the one for the f.c.c. structure shown in Ref. [136].

The step edge is a preferred adsorption site due to the lower coordination number of the step-edge atoms, and ad-atoms tend to accumulate at these sites (step-edge decoration). However, the diffusion length of the ad-atom on the terrace has to be larger than the terrace width, otherwise nucleation on the terrace will compete with step-edge decoration. The terrace itself is considered as a source of material, which is collected at the step edge. The control of terrace widths through the miscut therefore allows a homogeneous step-edge decoration to develop, although these step-edge wires tend to exhibit thermal fluctuations due to the interaction with the ad-atoms on the terrace, which can be described as a 2D gas. It might be more appropriate for some systems to consider a step-edge wire as a dynamic structure, which is controlled by rate equations for attachment and detachment of step-edge wire atoms.

A few examples for this process are the synthesis of Co wires at the step edges of Cu(775), Ag wires on Pt(997), Fe on Cu(111) and Au on a wide range of miscut Si surfaces, which are derived by tilting the (111) plane [111–113, 117, 118, 124, 137]. Figure 3.11 shows STM images of Fe wires on the Cu (111) surface, which are grown by step-edge decoration. It can be seen that an increase in the coverage leads to a somewhat inhomogeneous expansion of the wire into the terrace while for this system nucleation on the free terrace is suppressed even at higher coverage. Monoatomic metal wires and other low-dimensional structures, which are grown on metal substrates are unfortunately often highly susceptible to alloying, and interdiffusion.

3.4.2
The Smallest Wire – Electronic Structure of Monoatomic Wires

The Au–Si(111) system introduced by Crain *et al.* is one of the most versatile 1D systems [111, 124, 137], and is an excellent example that can be used to discuss electronic band structure of 1D systems, inter- and intra-wire interactions. The Au chain is built into the Si surface and becomes part of a reconstruction observed on the vicinal Si(111) surface, and the strong structural connection and bonding to the Si lattice stabilizes the 1D structure. This stabilization freezes the wire structure and prevents a Peierls distortion [125] of the chain atoms, which is predicted (and often observed) for metallic 1D systems. The Peierls distortion decreases the symmetry of the metallic system by readjusting the atomic positions and thus reduces the size of the Brillouin zone and opens a band gap. The overall energy of the system is lowered as a consequence of the distortion, but we no longer have a 1D metallic system.

The unit cell contains the Au-atom chain and the Si-surface atoms are bonded to the Au chain as shown in Figure 3.12 for the Au–Si(553) surface. This figure shows the band structure $E(k)$ as it was measured with angle resolve photoelectron spectroscopy with UV light, which probes the valence band (see Chapter 2), the experimental and theoretical Fermi surfaces. The 1D band structure follows the free electron dispersion curve of E–k_x where k_x is the direction along the chain, with additional bands below -1 eV stemming from the Si–Au bonded atoms, and Si bulk and surface atoms, which are part of the unit cell but not integrated in the 1D [124] Au-chain structure. The Fermi surface itself is a line, which confirms the 1D nature of the electronic system in the Au chain. The slight deformation in the Fermi surface is due to small contributions of 2D interactions with the Si surface, and with adjacent Au chains. 2D systems can be observed for some Au- and Ag-induced reconstructions in Si(111) and lead to a nearly perfectly circular Fermi surface [124]. Au on Si(111) is an excellent system to systematically observe the 1D to 2D transition and study the physics of low-dimensional systems.

3.5
Controlling the Electronic Properties of Semiconductor Nanowires

In Section 3.4 we focused on metallic nanowires extending our discussion of the electronic structure to the limit of monoatomic wires, which present truly 1D systems. The one-dimensionality was confirmed by measurement of the shape of the Fermi surface, which is a line for the 1D system and transforms to a circle for a 2D metallic surface. We now turn our attention

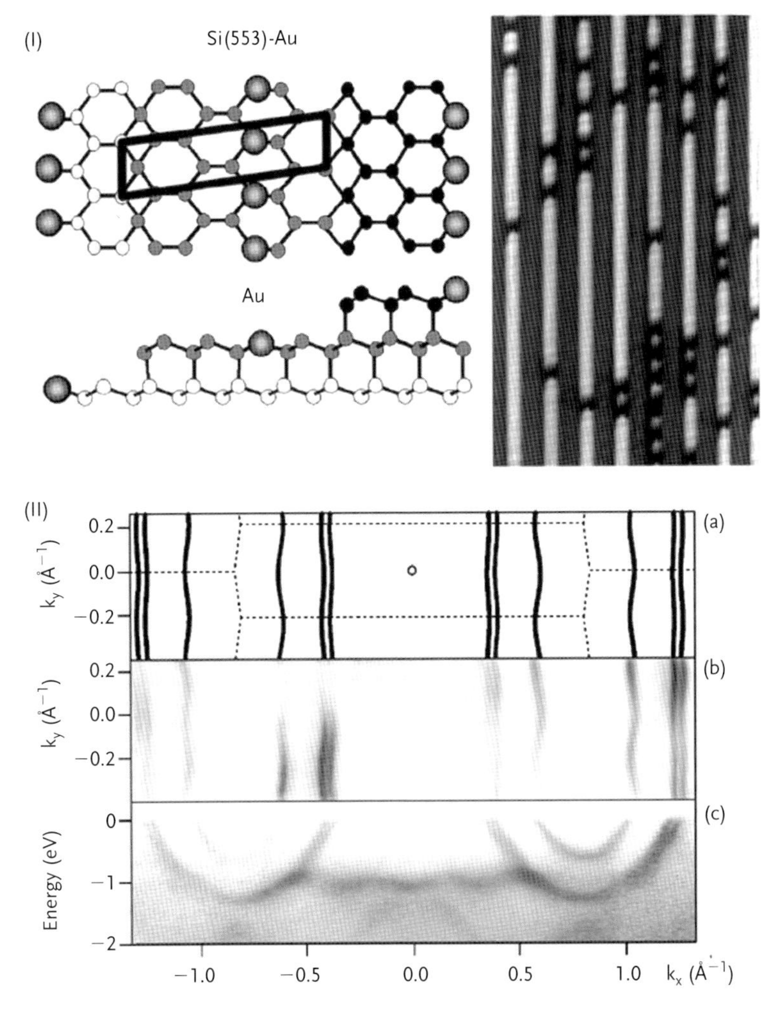

Figure 3.12 Band structure and Fermi surface of a 1D structure [137], which is designed by using a Si(553) surface. (I) Au deposition and annealing leads to the formation of Au chains and an STM image and the corresponding structural model are shown. (II) panel (c) Band structure ($E(k)$) as it was measured with angle resolved photoelectron spectroscopy (ARUPS). The bottom section of the $E(k)$ curves shows additional bands below 1 eV from the Si substrate. (II) panel (b) The Fermi surface which is shown in this panel is infinite in the k_y direction, with discrete lines in the k_x direction along the wires. Note: the panel (c) shows E as a function of k_x, in panels (b) and (a) the axes are k_x and k_y. In (II) panel (a), above the experimental data is the theoretical calculation of the Fermi surface shape and it is superimposed on the Brillouin zone of the Au chains (broken line). The deformation in the Fermi surface is presumably due to a small deviation from the perfect 1D system. Reprinted figure with permission from Physical Review Letters [137]. Copyright 2003 by the American Physical Society.

once more to the vertical, semiconductor nanowires, and discuss strategies which can be used to control and modify their electronic properties. We distinguish between two main pathways to electronic structure modification: firstly, confinement which relies on the control of wire diameter and is specific to nanostructures, and secondly, doping, which is well-known for all semiconductor materials but offers several surprising challenges in nanowires. Many of these challenges are also encountered in the doping of semiconductor quantum dots, since they are directly related to the small size of the system.

Traditionally, the manipulation of the electronic properties in semiconductor materials is achieved by doping, which introduces donor or acceptor states in the band gap in a controlled manner. Successful doping requires an intimate understanding of bonding, and position of dopant atoms within the lattice, and of acceptor or donor energy level with respect to the band edges. The latter determines the activation energy of the dopant and therefore its degree of ionization at a given temperature. Introductory lectures in electronic properties of materials often leave us with the impression that the choice of dopant is only governed by its ability to accept or donate charge to the semiconductor matrix, but often fail to discuss the equally important structural aspects. These include bonding at the correct site (interstitially and substitutionally bonded dopant atom can interact differently with the host lattice); spatial distribution of the dopant (cluster formation counteracts dopant efficiency); limited solubility in the host lattice; and diffusion and segregation of the dopant. For example, n-doping of diamond, which was highly sought after to make diamond a viable large band gap electronic material, has been hampered by many electronic and structural issues. These are to a large extent related to the short bond lengths in the diamond lattice, which prohibit the incorporation of many dopants. While p-doping of diamond with boron is fairly straightforward, the lack of a reliable n-type dopant has proven detrimental to the advancement of diamond electronics.

3.5.1 Controlling the Electronic Properties of Nanowires – Confinement

The challenges which are faced in adjusting the electronic structure and doping of semiconductor nanowires, and other inorganic nanostructures, arise directly from their small size and the large surface-to-volume ratio. The electronic structure of the smallest wires will be controlled by quantum confinement, which leads to a strong dependence of the band gap on wire diameter. This relation was demonstrated by an scanning tunneling spectroscopy (STS) measurement [138] of hydrogenated Si wires grown with an oxide assisted growth method, which allowed production of ultrasmall Si wires with a diameter down to about 1 nm.

The hydrogenated surface is essential in this experiment since it protects the wire surface against oxidation, and removes any surface states, which can arise from reconstruction on the side facets. The surface states are not only important for this particular measurement of the band gap but play an important role in the control of electronic properties of nanowires. Surface reconstruction leads, on the low index surfaces of Si, to the formation of extended surface states, which are part of the 2D band structure of the reconstructed surface. A surface state, which is positioned energetically in the band gap of the three-dimensional (3D) bulk-band structure can effectively screen the bulk bands and any dopant-related charges. The Fermi level is then pinned by these surface states, and any STS measurement will consequently measure the surface band gap and not the band gap of the "bulk" nanowire. The hydrogenation of the side facets therefore gives access to the bulk bands by removing the surface states of the facet.

Figure 3.13 compares theoretical [139, 140] and experimental results for hydrogen-terminated Si nanowires and illustrates the strong dependence of

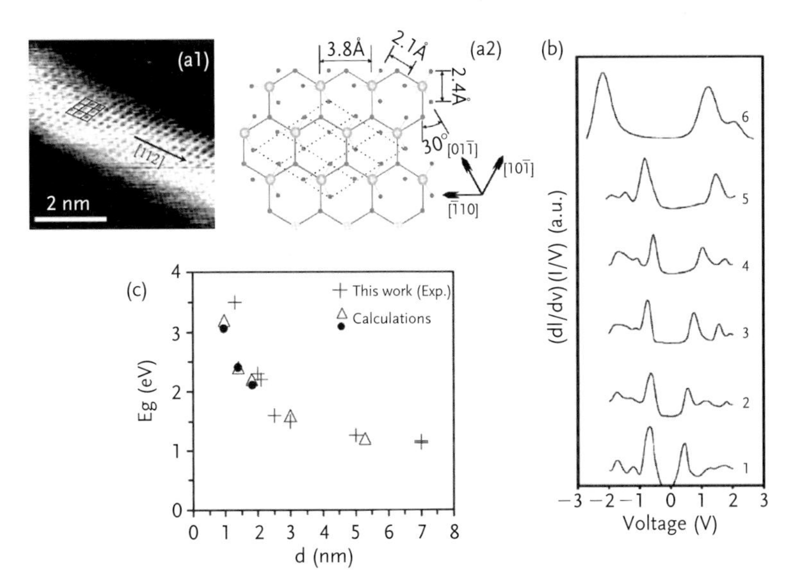

Figure 3.13 Si nanowires with hydrogen-saturated surface bonds [138] can be used to measure the band gap of the nanowire as a function of diameter with STM/STS. (a1) is an atomically resolved image of a (111) sidewall facet, and (a2) is the corresponding structure model; (the hydrogen are small dark dots which only have one bond to adjacent atoms, all other circles and dots are Si atoms). The Si nanowires grow predominantly with their axis in the [112] direction during the oxide-assisted growth process. (b) shows the differential conduction spectra recorded for five wires grown in [112] direction and one wire (number 6), which grew along the [111] direction. The band gap of the wires can be easily identified as the horizontal section at the center of the dI/dV curves. (c) The result of these measurements are summarized: the band gap is depicted as a function of wire diameter. The datapoints labeled "calculations" are from [139, 140]. From [138]. Reprinted with permission from AAAS.

band gap on wire diameter. The magnitude of the band gap can be used as a signature of the overall changes in the band structure with wire diameter. The Si wire band gap moves very close to the bulk value at a diameter around 8 nm, which is close to silicon's de Broglie wavelength of about 12 nm, which was introduced as a critical length scale for the onset of confinement in Chapter 1. The calculation of the full band structure for nanowires requires the same theoretical tools which are well-established for bulk materials, but it is possible to develop at least a qualitative understanding of some aspects of the size-dependent changes from the quantum mechanical description of the electron wave in a quantum well (particle in a box). This aspect is developed in Chapter 1, where the transition from a discrete energy spectrum to the quasi-continuum in a quantum well is discussed.

Quantum confinement not only affects the magnitude of the band gap but can also modify other aspects of the band structure. In Si the effective mass is highly directional, and for the six equivalent conduction band minima close to the X-point in the Brillouin zone of bulk Si the longitudinal effective mass m_L is 0.9163 m_0, while the transverse effective mass m_T is only 0.1905 m_0 [140]. The direction of confinement with respect to the direction of the conduction band minimum in turn defines the effective mass and relative weight of transverse and longitudinal components. The conduction band minimum with a small effective mass with respect to the direction of confinement will therefore be more strongly affected, and the position of the energy levels is shifted by a larger amount [140]. For Si the anisotropy in effective mass is quite large and the up-ward shift of the conduction band minimum due to confinement differs substantially for the different directions. In the case of Si this can even lead to the transformation from an indirect band gap of bulk Si into a direct band gap for Si nanostructures. The crossover to a direct band gap material due to size reduction was first recognized in the modeling of the astonishing properties of porous Si in the 1990s, which can be understood as a network of Si nanowires [139–141]. The transition from an indirect to a direct band gap material was first recognized in the optical properties of porous silicon, which shows a strong luminescence. This example illustrates that the general statement made earlier in this chapter, namely that Si is not a suitable material for optical applications, can be reconsidered in the case of Si nanostructures. The argument, which is made here, is certainly an oversimplification, but it served well to illustrate the complex role confinement plays in the modulation of electronic properties and bandstructure of nanomaterials.

In summary, for all common semiconductor materials used in nanowires, quantum confinement begins to play a significant role in the electronic transport and optical properties for diameters below the material specific de Broglie wavelength for the relevant charge carrier. The band structures of nanowires whose diameter is significantly larger than the critical diameter for quantum confinement will be equivalent or very close to the corresponding bulk material.

3.5.2
Controlling the Electronic Properties of Nanowires – Doping

The electronic properties of nanowires can also be adapted through the introduction of dopants in analogy to the modification of the corresponding bulk materials. However, the large surface-to-volume ratio and small dimensions of the wires introduce several additional challenges, which must be considered in the control of the electronic performance of the nanowire. In order to facilitate the discussion we distinguish between structural aspects, which includes the incorporation and segregation of dopant atoms within the wire, and electronic aspects, which are related to the modulation of the ionization energies, dielectric mismatch, and the presence of surface traps.

The doping of nanowires can be achieved by the incorporation of a dopant during the growth process, for example by adding a suitable gas-phase precursor. Ideally the dopant precursor is added to the wire by dissolution in the catalyst and consequently incorporation at the growth front. The dopant integration is subject to the same thermodynamic and kinetic considerations, which were discussed with respect to wire growth at the beginning of this chapter. However, due to the small energy differences in the nucleation of the WZ and ZB phases the addition of a dopant can also trigger a structural transition and substantially modify the parameter space for wire growth. This has been exploited by Algra *et al.* [91] to design twin superlattices in InP wires through a control of the Zn-dopant concentration.

The incorporation of a dopant will be influenced substantially by surface segregation, which is a well-known phenomenon for any kind of internal and external surface but becomes even more important in a system with large surface–volume ratio. Segregation is the "separation" of the bulk constituents, where one of the constituents moves to the surface or interface, which then becomes enriched with one of the bulk constituents (or in our case enriched with the dopant). Segregation is driven by the minimization of the overall free energy of the system, and is consequently material- and dopant- specific. While it does not affect the bulk properties of a doped semiconductor, given that the bulk semiconductor volume is sufficiently large, it will change the local concentration of dopants at each surface and interface. A change in the concentration profile of a dopant modifies the position of E_F in the gap, which is determined by the dopant energy level and concentration. As a consequence of the E_F shift, electron/hole transport across junctions will be changed and thus device performance is modified. A bulk semiconductor can be treated as a (nearly) infinite source of dopant atoms, and the bulk doping levels can therefore be retained. However, this is not the case for nanowires: segregation of dopants to the surface or, in the case of nanowires, the side facets, effectively removes the dopants from the bulk conduction channel, and thus reduces the doping efficiency. This can lead to a resistivity, which is close to that of the intrinsic material as shown for Si nanowires despite the attempt to incorporate a sizeable amount of dopant atoms during growth [54, 142–145]. Even relatively large wires in the diameter range above 50 nm can still suffer from segregation-induced dopant depletion and therefore present an increased resistivity compared to similarly doped bulk material.

However, dopant atoms located at the surface might be used to form a conducting surface channel, although this is only possible if the surface dopant can be ionized and thus participate in conduction. The activation of a dopant atom located at the surface is hindered by the modification of the ionization energies through dielectric mismatch (see below), and changes in the local bonding environment. Boron, for example, can occupy a surface site with a fully saturated set of three bonds to the neighboring Si atoms, and will therefore not donate a hole to a fourth bonding partner, which is missing at the surface site. However, while it is generally stated that dopant atoms at the wire surface are inactive, it might be possible to produce nanowires where the dopant atoms are located in the vicinity of the surface, and thus form a thin "cylinder" of dopant-enriched, electronically-active material around a dopant-depleted core with high resistivity.

Fermi level pinning [146] at the surface is a phenomenon which has been studied extensively due to its significance in the fabrication of semiconductor devices and junctions. The position of E_F with respect to the band edges is controlled by the charge carrier concentrations. If the concentration of traps and electronic states at the surface now differs substantially from those in the bulk, a potential difference develops which leads to band bending. The surface charge can arise from dangling bonds, defects, electronic 2D surface states, an oxide layer, or a compositional change due to segregation of dopants as discussed above. The band bending reaches relatively far into the semiconductor and its extension can be calculated from a self-consistent solution of the Poisson equation, which links the charge distribution, charge density and dielectric constant to the potential as a function of distance from the surface. The extension of band bending and therefore thickness of the charge depletion layer is material specific but can reach several ten nanometers and then controls the position of E_F throughout the entire nanowire.

Figure 3.14 [147] illustrates this effect: the transport characteristics and thus electronic function of the nanowire building block depends critically on the extension of the charge depletion layer, which can completely block the central transport channel even if this channel is not depleted of dopants.

The concentration of charge and traps on the surface can be very difficult to control. The behavior of the nanowire within a device can therefore become unpredictable, which leads to large fluctuations in device performance. Termination of the side facet surfaces and saturation of dangling bonds and defects states can promote the desirable flat-band situation throughout the wire. One example, which was discussed previously in the context of the

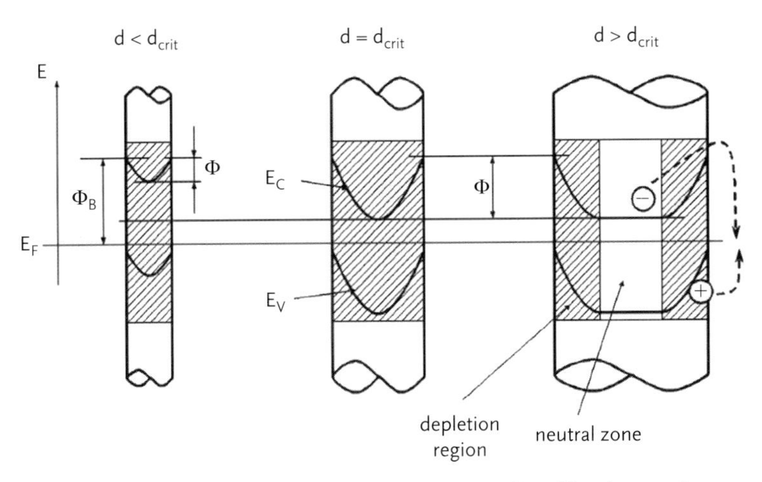

Figure 3.14 Schematic (not to scale) illustration [147] of band bending inside a semiconductor wire if the surface Fermi level (E_F) is pinned by surface states or traps. ϕ is the potential drop between wire surface and wire center, ϕ_B denotes the band gap, the energy difference between the conduction and valence band. If the wire diameter exceeds the dimensions of the band bending regions (d_{crit}) a neutral zone will develop at the center of the wire. The role of band bending on the wire performance was probed by Calarco *et al.* [147] through the observation of photoconductivity in GaN wires, and the charges on the right hand side indicate a surface recombination process, which follows the photoexcitation of an electron. Reprinted with permission from Nanoletters [147]. Copyright (2005) American Chemical Society.

measurement of Si nanowire band gaps as a function of diameter, is the hydrogen termination of Si facets, which removes the surface state due to Si surface reconstruction (Section 3.5.1). The formation of a co-axial wire system, where the core-wire is overgrown with a carefully selected thin shell is another approach to tailor doping, Fermi level position, and band bending within the nanowire system.

The last of the electronic effects, which we will discuss here, is dielectric mismatch. The dielectric mismatch refers to the abrupt change in the dielectric constant at the wire surface, and strongly affects the screening of the Coulomb potential of the ionized dopant. An ionized dopant can be described, following standard textbooks, as a charge orbiting an oppositely charged ion (exciton) in analogy to the hydrogen atom. The charge of the ion is strongly screened by the solid, and the Bohr radius of the exciton is at most a few nanometers [7, 148]. Dielectric mismatch becomes significant if the wire radius approaches the exciton radius and the Coulomb potential of the ionized dopant cannot be fully screened in the solid. The exciton radius is then compressed thus increasing its binding energy, and consequently the ionization energy of a dopant [149]. The Bohr radius of the first excited state, where an electron is excited across the gap into the lowest state of the conduction band, is also often used as the critical length scale for the onset of confinement.

3.6
Closing Remarks

Nanowires have recently been developed into highly versatile building blocks for nanoelectronics, photovoltaics, and sensing applications. A large number of fundamental studies on the growth process, and structure–property relationships now form the basis for the design of increasingly complex structures. Shape control is achieved by modulating catalyst diffusion at wire sidewalls, polytypism and crystallography are controlled through doping and cycling of growth conditions, branched structures are achieved by a second seeding of catalyst particles, and vertical and core-shell modulation of composition allows creation of quantum dots, and heterostructures on a single wire. Even the fundamental limit of a 1D nanostructure, where the wires are merely a single atom wide, has been studied, and device structures based on the manipulation of single atoms are envisaged. The next chapters will lead us to 0D structures, metal clusters and semiconductor quantum dots, and many of the ideas developed in this chapter, such as surface templates and band gap reduction with size, will be encountered again.

4
Metal Clusters

Metal clusters are ultrasmall metal particles, which occupy the transition regime between the single atom and bulk material. They are ubiquitous in nanotechnology applications, and have been studied for many decades because they are important in numerous catalytic processes. Metal clusters containing up to several hundred atoms have also been termed artificial atoms, since the electronic structure can be described in a similar manner as the orbitals in a single atom using a spherical Coulomb potential [151–164]. The valence electrons from the atoms (e.g., the s-electron from Na) occupy the cluster's energy levels. The nomenclature which is frequently adopted to denote the energy levels in a cluster therefore follows the chemistry notation of HOMO for the highest occupied molecular orbital, LUMO for the lowest unoccupied molecular orbital, and the electron shells are labeled with the quantum numbers s, p, d, f. . . . The metal clusters can be understood reasonably well using a spherical jellium model, which assumes a homogeneous distribution of background charge and delocalization valence electrons within the cluster. The properties, stability, electronic, and geometric structure of metal clusters fluctuate rapidly with the number of atoms N in the cluster.

The applications of metal clusters and small particles has recently expanded rapidly and they are now used, for example, in optical applications, where the surface plasmon is used as "guide" for the electromagnetic wave [165–167], tagging of biological systems, as vehicles for drug delivery [168], and nanoelectronic devices. One of the oldest historical examples for the use of Au nanoclusters is the glass–Au cluster composite; it was extensively used in medieval times to create many of the beautiful colors observed in church windows. The absorption spectrum of Au clusters is dominated by the plasmon resonance, which is the collective excitation of the cluster's free electrons, and its frequency and thus color of the clusters depends strongly on the cluster size. However, a complete treatment of the interaction of these composite materials with light is beyond the scope of this book.

Inorganic Nanostructures: Properties and Characterization, First Edition. Reinke, P.

Published 2012 by WILEY-VCH Verlag GmbH & Co. KGaA

The critical length scale for the electronic system in metal clusters is only a few nanometers, as discussed in Chapter 1, and larger clusters will behave electronically like the bulk material. This means that the electronic structure of metal clusters with fewer than several hundred atoms changes very rapidly. When we approach the regime of less than about 100 atoms (roughly 1.5–2.0 nm in diameter) the cluster structure can be modified by the removal of a single atom, thus placing stringent conditions on synthesis. For larger clusters with diameters between about 10 and 100 nm the electronic structure will be identical to that of the bulk material. These clusters still have a rather high percentage of surface atoms, and provide numerous step-edge and kink sites with high reactivity. Whether the increased reactivity of nanosize metal clusters is due to electronic or structural effects has been a long-standing question, which has been conclusively answered for only few systems [169–173].

The critical length scales for electronic and geometric structure therefore differ considerably: for very small clusters with relatively few atoms the electronic and geometric structures change rapidly, for larger clusters the electronic structure will be stable and bulk-like but the geometric structure and reactivity is still modified. An interesting example in this context is the melting point depression for small metal particles, which is due to the reduction in cohesive energy through the increasing contributions of under-coordinated surface atoms for decreasing cluster size. This effect is also discussed in the context of the vapor–liquid–solid (VLS) method for semiconductor wire growth, where metal clusters serve as catalysts.

4.1 Cluster–Surface Interaction

In Chapter 3 on semiconductor nanowires we distinguished between vertical and horizontal nanowires, and the interaction with the substrate was a decisive parameter with respect to the wire crystallography and growth. For small metal clusters we distinguish between free clusters in the gas phase, and clusters, which are in contact with a surface. The interaction with the surface modifies cluster shape, structure, and reactivity, and the latter is an important aspect in the design of catalysts. The surface is part of the reaction cycle and a synergistic interaction with the substrate is often observed and even desired. The substrate, which is most commonly an oxide support such as titanium dioxide or cerium oxide, can promote the overall catalytic activity. A catalytic process can then only be understood if the metal cluster and oxide support are treated as a whole rather than considered separately. Many aspects of substrate–catalyst interaction are hotly debated [3, 169, 172–176] and include electronic interactions, charge transfer and the stabilization of intermediate, metastable and highly reactive precursors.

While the discussion of catalysis itself is beyond the scope of our book, it is important to realize that clusters, which are supported by a substrate, or which are terminated by organic molecules for functionalization, can behave very differently from the free, gas-phase clusters. The interaction of metal clusters with metal surfaces or semiconductors is particularly strong and can lead to complete destruction of the cluster. The interplay between the cohesive energy of the clusters and the strength of the metal–substrate interaction determines the fate of a metal cluster as it is absorbed on a surface. However, several methods of growing and stabilizing metal clusters rely on a specific aspect of the metal–substrate interaction, such as the use of surface templates to define metal cluster arrays, which is analogous to the formation of horizontal nanowires discussed in Chapter 3. Two of many examples for templating on structured surfaces, which are discussed in more detail in Section 4.2.4, are Ir cluster arrays on graphene [177] (grown on Ir substrates), and ultrasmall clusters on the Si(111)-(7 × 7) reconstructed surface [178, 179].

4.2 Synthesis of Metal Clusters

The properties of metal clusters change rapidly with size, which places considerable demands on their synthesis. The goal is to achieve simultaneous control of size and spatial distribution (positioning) with a high yield and throughput. A brief overview of some of the commonly used methods is given in the following section: (a1) metal deposition – non-wetting; (a2) de-wetting of thin metal films; (b) gas-phase cluster sources; (c) aerosol-based formation of metal clusters; (d) co-polymer-assisted synthesis, which is solution based and also used to produce the highly stable Au_{55} clusters; (e) other solution-based methods; and (f) surface templates. Aerosols are included here because of their potential for industrial use, and application in the production of well-defined catalyst particles for the VLS growth of semiconductor nanowires (Chapter 3).

4.2.1 Non-Wetting Metal Clusters

The nucleation and growth of nanosize clusters is readily achieved by metal deposition on a non-wetting substrate, where the thermodynamic equilibrium favors the formation of islands and clusters over layer-by-layer growth. The metal deposition is achieved by exposure of the substrate to a

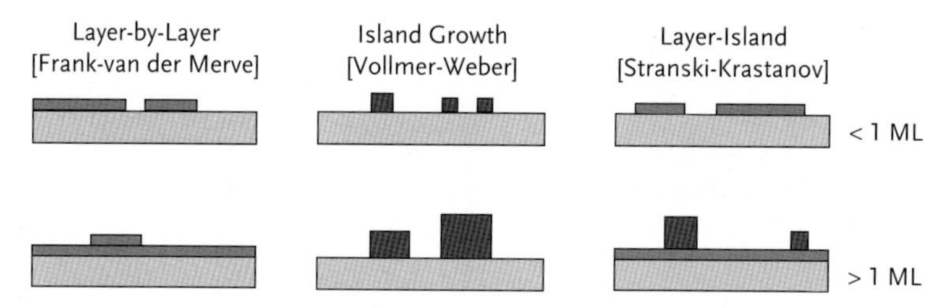

Figure 4.1 Fundamental growth modes for ad-layers on flat surfaces in thermodynamic equilibrium. The growth mode is determined by the respective surface and interfacial energies.

beam of thermal metal atoms, which are created by evaporation from a solid source such as a Knudsen cell, an electron beam evaporation source or a sputter source. The metal ad-atoms adsorb on the substrate surface and will then adopt a structure, which is dictated by the kinetic and thermodynamic conditions of the materials system (Chapter 1). For a system in thermodynamic equilibrium, the overlayer growth follows one of two modes (see Figure 4.1) layer-by-layer and wetting (Frank–van der Merwe) or island growth mode (Vollmer–Weber), which is the equivalent to non-wetting. The third growth mode (Stranski–Krastanov), where a layer-by-layer growth switches to an island growth mode at some point in the growth process, is only realized for selected systems such as a highly strained overlayers in the formation of quantum dots (see Chapter 5). The equilibrium of surface and interfacial energies between the solid, liquid, and vapor phases determines whether a wetting or non-wetting growth mode prevails. The contact angle Θ, which is the angle at the intersection between substrate surface and boundary plane of the liquid drop/overlayer–vapor/gas interface, is related to the surface energies by Young's equation:

$$\cos\Theta = \frac{\gamma_{\text{liquid-solid}} - \gamma_{\text{vapor-solid}}}{\gamma_{\text{vapor-liquid}}} \tag{4.1}$$

where γ denotes the surface energy at all interfaces/surfaces of the system as shown in Figure 4.2. While Young's equation is formulated for contact angles of liquid droplets the same energy considerations hold for solid–solid system. The liquid droplet–solid interaction has been identified as a critical aspect in the growth of semiconductor nanowires, where the droplet sits on top of the wire and the interface is at the same time the growth plane of the nanowire (Chapter 3).

Most metals will indeed not wet an oxide or semiconductor surface, which is an excellent starting point for the synthesis of small metal clusters. The cluster size distribution is usually rather broad and driven by the interplay between nucleation, growth and ripening. The position of the clusters

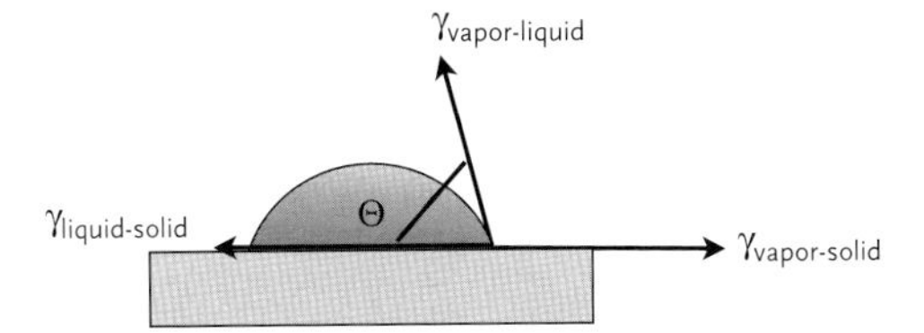

Figure 4.2 Liquid droplet in thermodynamic equilibrium; the respective surface and interface energies are indicated, and the contact angle Θ is determined by Young's equation (Equation 4.1). The case shown here is for a liquid which wets a solid surface.

is random and the average distance and size are controlled by the characteristic length scales of diffusion and transport on the substrate surface. On first sight the lack of control over the size distribution appears prohibitive for many nanoscale applications, but this is not necessarily the case. Nucleation and growth on a non-wetting substrate are quite versatile and ease of implementation can be a great advantage.

One example which illustrates the use of the simple "nucleation and growth" method for the fabrication of nanoscale electronic devices is the formation of a single electron transistor [180] (SET, see Chapter 7.1) from Au clusters deposited in the gap between two electrodes. The positioning of the clusters cannot be controlled and the characteristics of the SET depends on the size of the metal cluster, which is by chance formed in the electrode gap and in contact with both electrodes. However, while the SET characteristics change rapidly as a function of the metal cluster size, the reactivity of metal clusters often changes much less dramatically. A relatively broad size distribution, as long as it coincides with the regime of enhanced reactivity, is therefore not prohibitive. The additional gain in reactivity and yield, which would be achieved through a narrower size distribution, is outweighed by the cost of a more complex synthesis required to achieve size control.

4.2.2 Aerosols and Cluster Sources

Cluster and aerosol sources [181–187] allow the synthesis of gas-phase clusters with very narrow size distributions, which are not modified through the interaction with a surface. This means that the fundamental properties of metal (and insulator) clusters can be studied, including their geometric and electronic structure, which is discussed in detail in the next sections. The narrow size distribution of metal clusters, which are used as catalyst particles in the growth of nanowires (Chapter 3) or carbon nanotubes (Chapter 6) leads to a superb control of the nanowire and nanotube properties, due to the coupling of cluster size with wire diameter.

Aerosols are suspensions of solid particles in the gas phase, and they travel from a production zone to a charging device, which creates positive and negative ions, which is the prerequisite for the subsequent step of mass separation. The clusters are transported through the different zones by the drag exerted by a carrier gas such as N_2. The formation of metal clusters/particles in the production zone, which can be a hot tube furnace with a molten metal source, a laser ablation set-up, or a low pressure plasma, is achieved through a gas-phase nucleation and growth process. Once the metal clusters leave the production zone and are ionized they can be separated using, for example, a differential mass analyzer (DMA). Ionization and separation of aerosols are well-established techniques, which are used in many industrial processes. The width of the cluster size distribution is relatively broad at the exit of the production zone, but is very narrow when the cluster beam exits the DMA. A width of the size distribution of 5% of the mean cluster diameter has been cited [185, 186]. The clusters often have a more fractal shape upon exiting the production zone, and post-synthesis sintering is required to achieve the desired spherical shape.

The supersonic expansion from a cluster-seeded gas phase has emerged as another important method to synthesize clusters of many different metals and insulators with up to 10 000 atoms with a mass resolution down to a single atom [181, 182, 184, 188]. The gas-phase seeds are created in a relatively high-pressure environment with an inert background gas, which promotes and controls the collision rates and thus the size of the seed. The nucleation and growth process occurs within the gas phase, and the choice of carrier gas and pressure are critical for an efficient cluster production. Equilibrium size distributions are often achieved due to the very high collisions rates between clusters. The high-pressure beam is then expanded through a supersonic nozzle, which interrupts the cluster growth process and essentially preserves the size distribution acquired during cluster growth in the high-pressure zone. The number of gas-phase collisions is minimized after the expansion, and ripening will be suppressed. The subsequent ionization and separation with a mass spectrometer completes the synthesis, and it is possible to achieve a mass resolution sufficient to obtain a truly monodisperse size distribution and thus ultimate control of cluster size.

Measurement of the cluster size distribution after the expansion and prior to mass separation offered for the first time information about stability of small metal clusters because these distributions are observed on a cluster population reflecting thermodynamic equilibrium populations [158, 161]. The most striking observation is a series of "magic" cluster sizes which are much more abundant in the size distributions due to their exceptional stability. Magic clusters are observed in many different material systems, although their size depends on the specific mechanism responsible for their stabilization. The magic clusters are discussed in more detail in the next section following the description of the micelle method for cluster synthesis.

4.2.3 Synthesis and Stabilization of Metal Clusters

The co-polymer-assisted or micelle method is solution based, and one of the most versatile approaches to cluster synthesis and it is becoming increasingly important in the large-scale synthesis of metal clusters. Numerous derivatives of this method have been developed over the years, and they are used in a wide range of applications, and have led to the fabrication of wires, clusters, quantum dots, and many other nanostructure materials and geometries. The co-polymer method and related techniques are not only an elegant approach to the synthesis of clusters (and quantum dots) but also a pathway to the functionalization of cluster surfaces for medical or sensing applications by highly specific ligands. The advantages lie in their versatility, the ability to control cluster size and spatial arrangement, and the inhibition of cluster coalescence. However, the clusters are coated with an organic material (ligands), which can modify their properties [189] and is prohibitive for many application that rely on the specific electronic or geometric structure of the cluster surface.

Co-polymer-assisted synthesis [159, 189–192], which is only one example of the solution-based methods to nanomaterial fabrication, can be classified as a template-based bottom-up approach to the fabrication of nanostructures, and exploits the structural variability given by block co-polymer structures. (More solution based methods are discussed in chapter 5 on quantum dots) A co-block polymer consists of a set of two (or more) distinct polymer chains, which define the "blocks." These chains are connected through covalent bonds, and as a consequence the different polymers cannot de-mix and complete phase separation is prevented. The block co-polymers therefore often separate into microphases, with short length scales, which are defined by the relative length of the polymer chains. One type of microphase-separated block co-polymer are micelles: a hydrophobic and hydrophilic polymer chain (or chain-terminating group) are combined and in an aqueous solution the hydrophilic end will face the outside, while the hydrophobic ends face inward [192, 193]. The opposite orientation is achieved in an organic non-polar solvent. Micelles are formed, which are spherical particles with a hollow core, whose walls are made of the polymer chains.

A metal salt is then added to the solution and fills the hollow core of the micelle, where it is encapsulated and protected. Subsequent reduction of the metal yields a well-defined metal cluster whose size is defined by the structure of the micelle, and inter-cluster distances are controlled by the length of the polymer chains. The role of the organic layer in the

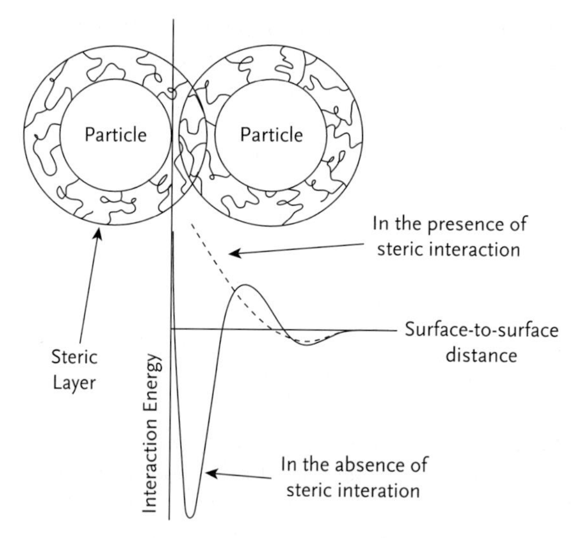

Figure 4.3 Schematic illustration [159] of the interaction potentials between two metal particles/clusters: (solid line) interaction potential for metal particles, which shows a deep attractive well when the particles come in contact, (broken line) interaction potential for two particles, which are coated with a steric layer. The steric layer can be a polymer coat, micelles, or organic groups tethered to the surface of the metal particle. A small attractive potential well is seen when the outer perimeters of the steric layers touch, but the potential is repulsive as they come closer and begin to interact. The presence of the organic layer leads to steric repulsion, and prevents the metal particles from touching and coarsening. Reprinted from J. Mol. Catal. A:Chem. [159], Copyright 1999, with permission from Elsevier.

spatial arrangement of clusters is illustrated in Figure 4.3: the interaction potential between metal clusters is modified by the steric constraint introduced with the organic layer, which prevents coarsening and therefore stabilizes the cluster size distribution. Dispersion of the micelles on a substrate can now be used to create a spatially ordered array of metal clusters with a homogeneous size distribution. The micelle can be removed through mild oxidation methods using oxygen or hydrogen plasma etching, leaving the naked metal cluster behind. Boyen *et al.* used this approach to study alloy formation in large, regular arrays of size selected Au clusters [50].

In addition to metal clusters nestled inside a micelle, it is also possible to attach organic ligands directly to the metal cluster, which allows metal clusters to be coupled in a well-defined manner to surfaces or the functionalization required for a variety of applications to be achieved. One of the most prominent examples of a ligand-stabilized metal cluster [159] is Au_{55}, which is terminated with twelve PPh_3, triphenylphosphine, and six Cl ligands making it $Au_{55}(PPH_3)_{12}Cl_6$. However, modification of the metal cluster properties through the attachment of a ligand is often not well understood and remains one of the major challenges in the use of ligand stabilized clusters. An experiment published by Boyen *et al.* [189] unravels the impact of the ligand in $Au_{55}(PPH_3)_{12}Cl_6$ on the electronic structure of the Au core. The presence of the Cl and PPh_3 ligands transforms the metal core into an insulator, and removal of the ligands either by a plasma treatment or radiation will switch the Au-core back to its metallic state. The Au_{55} clusters are just large enough to be metallic, while a band gap will open in smaller clusters as discussed in more detail in the next section.

4.2.4 Clusters on Surfaces: The Smallest Templates

The last part of this section on the synthesis of metal nanoclusters will discuss selected examples for surface-supported clusters, which bear some similarity to the horizontal wires discussed in Chapter 3. Surface-supported clusters form when the surface itself is non-wetting with respect to the cluster material, and we will now take a closer look at surface-supported clusters where the surface structure itself functions as a template. The surface template serves to stabilize certain cluster sizes, and it can be used to control the spatial distribution, which is often elusive with other methods. We will use examples from recent experimental studies to illustrate the role of a surface template in the deterministic fabrication of cluster arrays. These example include the formation of ultrasmall Al, and In clusters with only a few atoms on the reconstructed Si(111)-(7 $\times$ 7) surface [28, 34], and the growth of Ir clusters on graphene, which is itself grown on an Ir substrate [177].

There are numerous material systems where surface templates are used to stabilize metal clusters, and the two examples are chosen to illustrate the interplay between the surface geometry and electronic structure in driving the cluster assembly. For the step-edge decoration discussed in Chapter 3 a naturally occurring one-dimensional surface inhomogeneity, a step edge, is used as template for the formation of horizontal wires – in analogy, for the synthesis of zero-dimensional (0D) clusters it is necessary to provide an array of 0D "attachment points," which act as nucleation centers for clusters. An "attachment point" is a minimum in the potential energy surface, which represents the interaction between adsorbate and surface. The attachment sites will only fulfill their function as templates if the surface mobility and mean free path for ad-atom diffusion is larger than the distance between attachment sites allowing for selective nucleation of clusters at these preferred sites.

The first example for the surface-templated synthesis of nanocluster arrays are metal clusters on the Si(111)-(7 $\times$ 7) reconstructed surface. The surface template serves a twofold purpose: it creates a regular array of "attachment sites" where nanoclusters with well-defined sizes are formed, and it stabilizes an array with relatively small inter-cluster distances. The Si(111)-(7 × 7) surface, which functions as a template in this example, is one of the most famous surface reconstructions, and it was one of the first surfaces imaged with scanning tunneling microscopy (STM) by Binnig and Rohrer [28, 33]. The (7 × 7) reconstruction is shown schematically in Figure 4.4. It is a relatively complex reconstruction, which serves to minimize the number of Si surface atoms with dangling bonds. The reconstruction consists of two halves, which are structurally equivalent and separated by a mirror axis. Due to this symmetry element, which is not part of the cubic structure of the Si bulk, it is necessary to introduce a stacking fault between one half of the surface unit cell and the bulk lattice. This is the faulted half of the unit cell (FHUC) of the (7 × 7) reconstruction, the other half is the unfaulted half of the unit cell (UFHUC). The Si(111)-(7 × 7) reconstruction has been shown to act as a rather universal template for the assembly of nanocluster arrays for a wide range of metals, including but not limited to Ag, In, Mn, Tl, Na, and Pb [194–196].

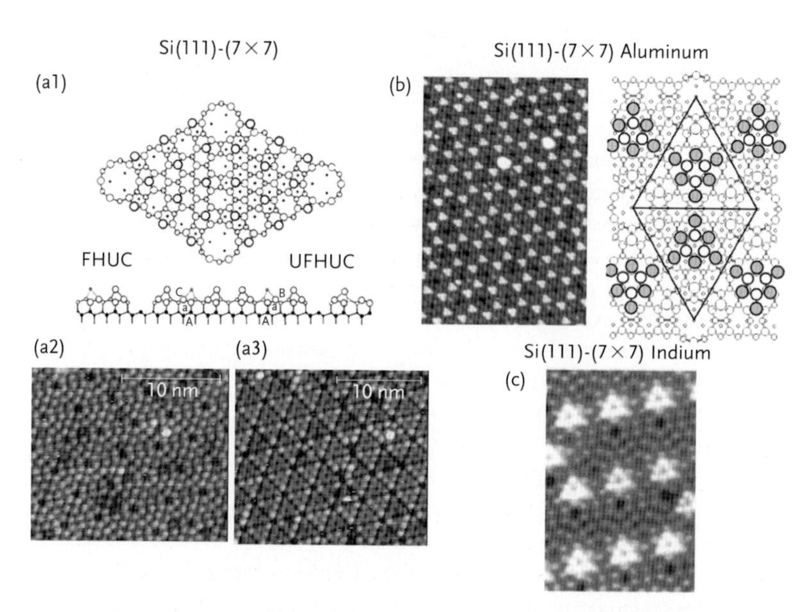

Figure 4.4 STM images and models of the surface reconstruction of Si(111)-(7 × 7) [34]: (a1) includes top and side view, which shows the stacking fault in the right half of the reconstruction unit cell, (a2) STM image probing the empty states (0.03 nA and bias voltage of 1.5 V), and (a3) filled-state image of the same surface (0.03 nA and −1.5 V). Images courtesy of K.R. Simov, and G. Ramalingam. On the right-hand side: (b) Al clusters [178] (0.35 ml) on the Si(111)-(7 × 7) [V_{bias} = 2.0 V] and the corresponding structure model, and (c) In clusters (0.05 ML) [195] [V_{bias} = 0.6 V]. (a1) Reprinted with permission from [34]. Copyright 1985, American Vacuum Society. (b) and (c) Reprinted figures with permission from Phys. Rev. B and Phys. Rev. Lett. [178, 195]. Copyright 2002 by the American Physical Society.

Figure 4.4 includes STM images and models of the Si(111)-(7 × 7) reconstruction, the Si(111)-(7 × 7) with In and Al clusters [178, 195]. The deposition of In leads first to the formation of clusters consisting of six atoms, which are exclusively formed on the FHUC of the unit cell. The In atoms are captured at the center of the FHUC, and are thought to be highly mobile within that area but unable to leave their preferred position. The structure of the clusters is clearly driven by the interaction with the surface. The edge of the FHUC, which is formed by differently bonded Si atoms, is an efficient barrier, which cannot be overcome at room temperature and it serves to stabilize the ultrasmall clusters. This barrier at the same time limits the attainable cluster size to a few atoms, and the intercluster distance is about 2.7 nm. The stability of the In clusters is quite exceptional and they only start to dissolve at about 200 °C. Once all FHUCs are filled, the clusters start to form on the UFHUC part of the unit cell. The Al nanoclusters, which are also shown in Figure 4.4 do not appear to have a preference for the FHUC or UFHUC. A large number of reconstructions on different surfaces have been used as templates for nanoclusters and nanowires, although they are limited to the lower end of the cluster size range. One challenge, which must be overcome in order to fabricate surface-reconstruction templates for any kind of application, is the sensitivity of these surfaces to oxidation. Capping of nanostructures while retaining their coveted properties such as magnetism, or single electron transport, will become increasingly important in future nanoelectronics devices.

The second example for surface-templating is the formation of a regular array of Ir clusters, which is self-assembled on a graphene layer grown on an Ir(111) substrate [177]. Figure 4.5 shows the very regular cluster arrays, which are formed on the Ir–graphene surface. In this case the attachment centers are not created by a surface reconstruction but by the distortion of the surface due to the lattice mismatch between the Ir substrate and the graphene layer. The Ir surface structure is not commensurate with the graphene's honeycomb lattice, and the slight misalignment between the two structures is expressed in the STM images in a Moiré pattern with periodically alternating bright and dark regions. The different local contrast corresponds to a difference in the alignment of the respective lattices and therefore a variation in the electronic and bonding interaction between Ir surface and graphene. The (9 × 9) Ir surface structure is matched with a (10 × 10) graphene mesh and this mismatch leads to a longer range Moiré type corrugation mitigated by local, periodic variation of graphene–Ir interaction. This creates an "electronic" superlattice whose periodicity is given by the wavelength of the Moiré pattern. The Ir clusters, which form on the graphene surface occupy preferential sites in the Moiré pattern – the attachment site is defined by substrate–graphene interaction. The cluster sizes are around 70 ± 10 atoms and are considerably larger than the ultrasmall clusters accessible through templating with surface reconstructions.

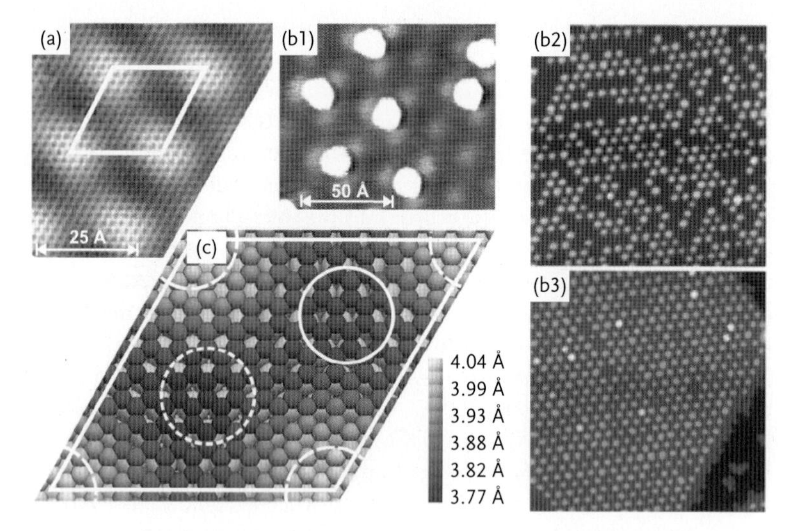

Figure 4.5 STM images and structural model of Ir clusters on an Ir–graphene substrate as described by N'Diyae *et al.* [177]. (a) Graphene layer on Ir with a Moiré structure. (b1) Ir clusters, which form at specific sites within in the Moiré structures after deposition of 0.02 ML of Ir, (b2) cluster array with 0.03 ML Ir, and (b3) cluster array after deposition of 0.1 ML Ir. The image sizes are 55^2 nm^2 and the imaging conditions are a bias voltage of 0.2 V and a feedback current between 8 and 23 nA. (c) shows the detailed structure of the Moiré unit cell obtained from DFT calculations: the solid line indicates regions with an h.c.p. type stacking between the Ir substrate and the graphene layer, the broken line denotes a region with f.c.c. type stacking. Reprinted figures with permission from Phys. Rev. Lett. [177]. Copyright 2006 by the American Physical Society.

4.3 Geometry of Clusters

In this section we will focus on the geometric and electronic structure of free, gas-phase metal clusters. We will start by looking at ultrasmall clusters, which are made of 2 to about 50 atoms in size, and whose structure and properties change quite dramatically with size. These clusters are dominated by surface atoms, which are defined as atoms with fewer nearest neighbors than expected from their respective bulk structure. Ultrasmall clusters truly represent the transition regime between the single atom and the bulk, both geometrically and electronically. The consideration of cluster stability and bonding configuration for ultrasmall and larger clusters up to at most a few thousand atoms leads us to the so-called magic clusters, which are specific cluster sizes, which exhibit exceptional stability. The ability to explain the presence of magic clusters serves as a yardstick for theoretical descriptions of cluster structure and properties. The fluctuations of specific properties such as band gap, melting temperature, and ionization energy with cluster size likewise reflect the modulation of cluster stability as a function of size.

4.3.1 Shells of Atoms

The question of cluster stability is closely connected to the question of cluster shape and how the shape evolves with the number of atoms within a cluster. The smallest possible cluster, which might be more appropriately called a molecule, is a two-atom unit. When more atoms are added to the molecule it will at some point transition to a spherical shape, where the number of fully coordinated atoms is maximized, and then begin to exhibit facets, which minimizes the number of kink and step sites on the surface and reduces strain.

Metal clusters will approach a spherical shape if the number of atoms approaches about 50 to 100 atoms, for smaller clusters a large number of under-coordinated atoms has to be accommodated [151–156, 158, 159, 161, 197, 198]. The cluster shape is dominated by the drive to simultaneously minimize the bond angle and bond length distortions around individual atoms, and maximize the number of bonding partners for all atoms within the cluster. This appears to be an easy task if one has only a few atoms to consider, but becomes increasingly difficult and computationally expensive for larger systems of 10 to 50 atoms. The immediately intuitive progression of cluster shape, which minimizes the number of surface atoms, leads to a sequence of polyhedra with hexagonal closed packing, as shown in Figure 4.6, where completion of an atomic shell will yield the most stable clusters. However, electronic contributions, and directional and covalent contributions to bonding lead to energetically favorable structures, which will not adhere to the simple rule of geometric shell addition [163]. In order to optimize the geometric structure of the clusters it is therefore necessary to probe a much larger number of possible geometries for a given number of atoms using additional boundary conditions. The number of possible structures rises rapidly with cluster size, and often more than one solution can be obtained, which correspond to different geometric structures with only small differences in stability. Several structures, so-called isomers, can therefore coexist for a cluster with N atoms.

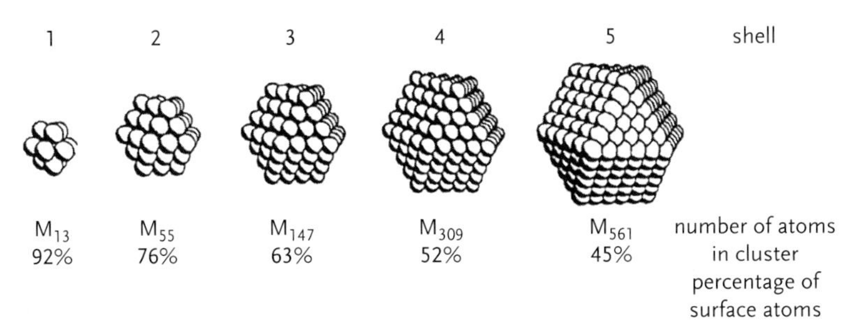

Figure 4.6 Geometric shell model, where the coordination number of atoms within the cluster is maximized using a hexagonal closed packing [159]. Reprinted from J. Mol. Catalysis A; Chem. [159]., Copyright 1999, with permission from Elsevier.

An example of cluster shape evolution for $N<55$ ($N=$ number of atoms in cluster) is the series of Au clusters shown in Figure 4.7. This sequence of structures was calculated using density functional theory (DFT) methods, which include the 5d and 6s electrons, and some relativistic contributions [154]. The clusters show a progression from a linear to a planar shape, which then transforms into a somewhat open three-dimensional (3D) structure, and for the largest N a closed sphere is obtained. The cluster shape approaches the ideal spherical shape for N around 50. The shape of these clusters cannot be described by the translational symmetry of a unit cell, which is the basis of understanding the bulk crystalline structure. They are therefore considered noncrystalline or amorphous. The progression from two- to three-dimensional cluster shapes appears to be rather universal, although the cluster size at which the transition occurs depends on the element, due to subtle differences in bonding.

Experimental determination of cluster shapes in the regime of ultrasmall clusters is quite challenging since immobilization on a surface often modifies the cluster shape and properties, therefore making them inaccessible to microscopy techniques such as TEM (transmission electron microscopy) or STM. One of the few measurements of the geometry of gas-phase clusters has been achieved with infrared spectroscopy [156], which probes the vibrational modes in gas-phase clusters. In combination with the simulation of cluster shapes this experiment identified cluster

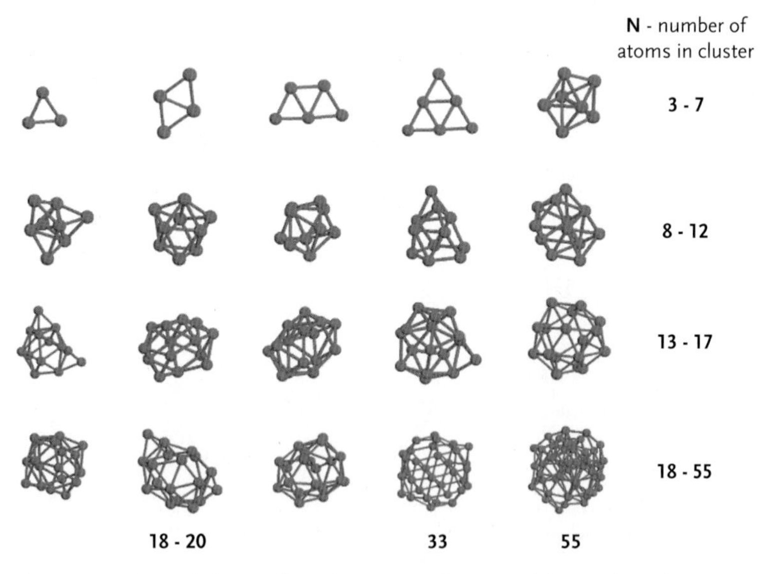

Figure 4.7 Shape of ultrasmall Au clusters as a function of the number of atoms. The larger clusters approach a spherical shape [154]. Reprinted with permission from [154]. Copyright 2007 American Chemical Society.

geometries. It was even possible to identify several isomers and clusters, which contain noble gas atoms attached during the synthesis.

Another example which illustrates the relationship between bonding and cluster geometry, is the study of binary alloy clusters [155, 199]. Figure 4.8 shows one example for a family of alloy clusters with an outer shape of a truncated icosahedron containing 38 atoms. The figure illustrates the variation in the positions of the two elements (A = Ag, Au and B = Cu, Ni, Pd, Co) in the cluster as a function of relative element concentration. The transition from a core-shell type to an intermixed cluster can easily be identified and occurs when the same number of atoms from both elements is used. Alloy clusters are of considerable technical importance:

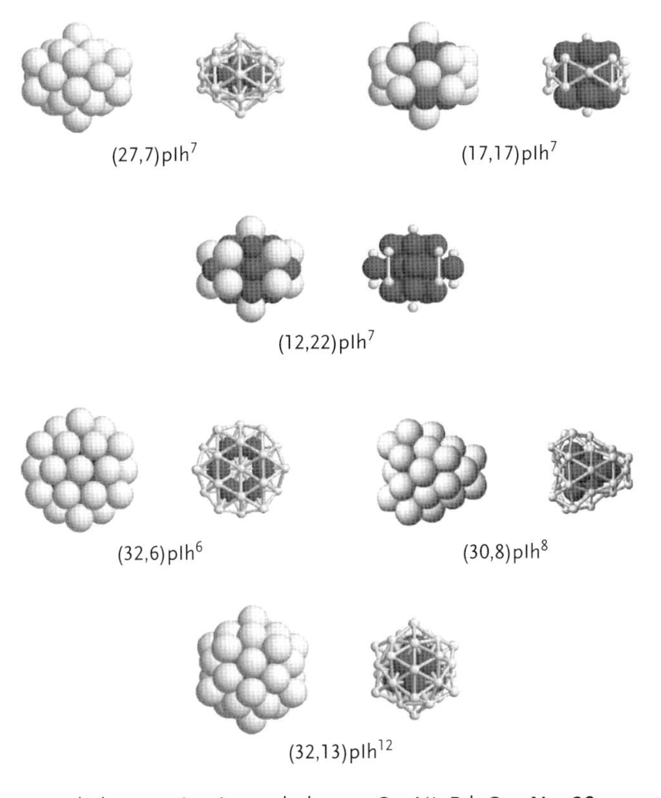

Figure 4.8 Ultrasmall binary alloy clusters with the geometry of a truncated icosahedron. The labels at each model denote the number of atoms type A (light grey, noble metal), and type B (dark grey, transition metal). The symmetry group of each cluster is determined by the relative arrangement of the elements and denoted for each model. Reprinted figure with permission from [155]. Copyright 2005 by the American Physical Society.

most industrial catalysts use expensive and rare noble metals, and it is important to search for cheaper materials which at the same time retain the function of the noble metal catalyst. One solution to this problem is to use particles, where only the highly reactive surface is made of the noble metal, while the inside of the particle is made of a more abundant and cheaper material. A well-defined core shell arrangement of the elements is in many cases considered to be the ideal geometry. This approach only works if the reactive surface is stable and can be maintained for extended periods of time. Alloys and core-shell clusters in nanoscale particles represent systems which are miscible or immiscible, respectively. Although, once again, the bulk phase diagram is only a poor predictor of nanoscale material behavior: some combinations of elements, which are immiscible in the bulk can indeed form fully miscible alloys [200], and vice versa.

4.3.2 Magic Clusters and Stability

The relative stability of metal clusters is measured by analyzing the abundance of cluster sizes within a cluster beam, which has reached thermodynamic equilibrium prior to supersonic expansion. The abundance spectrum of Na is shown in Figure 4.9 and it is readily apparent that certain cluster sizes are considerably more abundant than others [161, 201]. The larger abundance is due to the higher stability of these clusters, which were subsequently labeled "magic clusters" and coincides, for Na clusters, with the filling of electronic shells in analogy to the enhanced stability of noble gas atoms. Figure 4.9 includes a calculation of the electronic stability for electrons in a spherical potential well and the magic cluster sizes observed in the mass spectroscopy abundance spectra agree quite well with the occurrence of a filled electronic shell. The ability to describe the abundance spectra and recognize magic cluster sizes serves as a yardstick for theoretical models. Na is an excellent model system in this context and can be treated as contributing only free-electron-like s-electrons. The investigation of systems with d-electrons such as Cu, Ag, or Au is computationally challenging [160, 161, 202]. The stabilization of a specific cluster size can be due mostly to electronic contributions, as shown here for small Na clusters, or to geometric constraints, which was shown by Martin *et al.* [203] for large Na clusters. They reported fluctuations in cluster abundance for Na clusters with up to 20 000 atoms[1], and attribute the stability modulation in this size regime to geometric shell closure.

[1] A cluster with 20 000 atoms is still rather small and will have a radius around 3–4 nm (depending on packing density of atoms this number is only approximate).

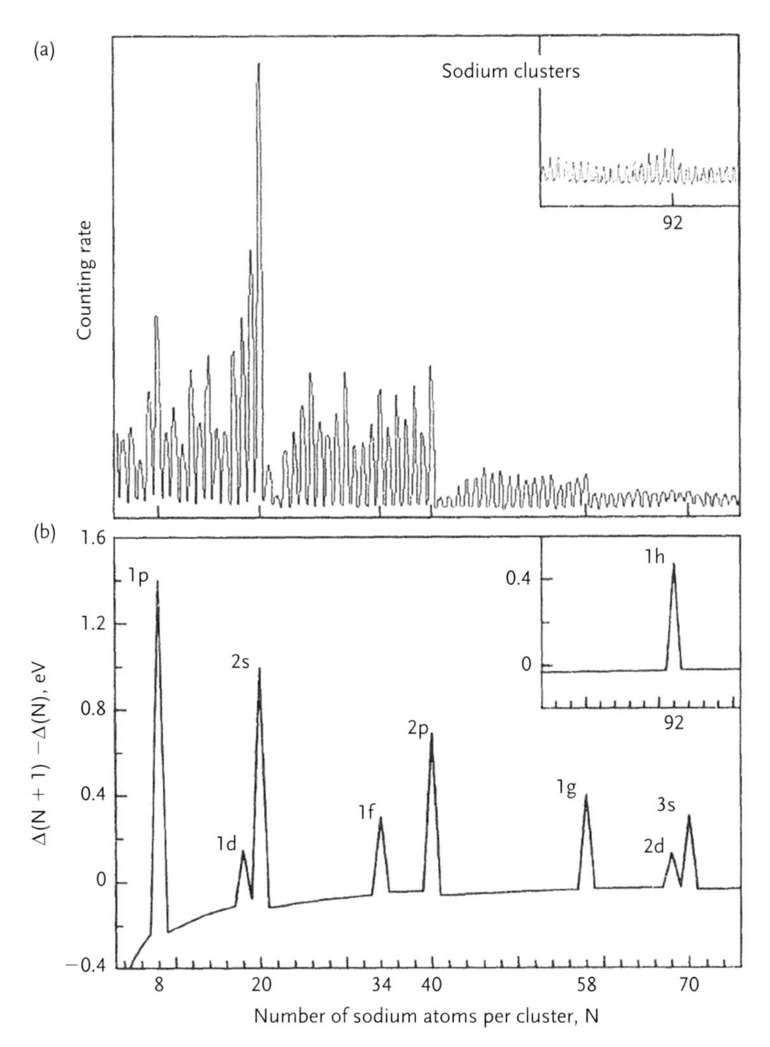

Figure 4.9 Abundance distribution of Na clusters, which are produced in a supersonic expansion cluster source [161]. The mass spectrum of the clusters is shown in (a) and the peaks with significantly higher counting rates are magic cluster sizes. (b) shows the result of an energy level calculation for a spherical potential well. The change in stability is plotted with respect to the number of atoms per cluster and the peaks coincide with complete filling of an electron shell. The peak nomenclature is in analogy to the energy levels in atoms. The agreement of the position of enhanced electronic stability and magic cluster size is an indication that electronic shell filling is a significant contribution to the overall cluster stability. Reprinted figure with permission from [161]. Copyright 1984 by the American Physical Society.

For small Na clusters the shell model and the jellium model which focus on the electronic contributions, have been particularly successful in understanding the magic cluster distribution. The shell model was developed in analogy to the shell model used for the description of the nucleus but is also closely related to the quantum mechanical description of the electronic levels in a single atom. It is assumed that each Na atom donates a single electron to the cluster, which then is treated as a spherical potential well in the shell model. The potential is parameterized and thus a relatively simple solution to a many-body problem can be used. The solution of the Schrödinger equation then delivers a set of energy levels, which correspond to shells and thus quantum numbers in the clusters. A completely filled shell is particularly stable. The shell sequence is 2, 8, 20, 34, 40, 58, . . . for the spherical well.

The jellium model includes a self-consistent description of the electronic system and assumes a featureless background of ionic charge. However, it can be adapted to include different cluster structures, which deviate from the spherical shape. Details on the computational aspects of the different models are given in Ref. [152]. Both methods are quite accurate in their description of the abundance spectra of Na atoms, and confirm the role of electronic stabilization of metal clusters. A comparison of cluster abundance spectra, and cluster geometries for different elements is given in Ref. [158].

While this chapter has been focused firmly on metal clusters it is very informative to take a closer look at the cluster abundance spectra for TiN. TiN is a very hard material with metallic conductivity due to its unique bonding configuration with ionic (the Pauling electronegativity difference is 1.5 between Ti and N), covalent, and metallic contributions. It crystallizes in a cubic NaCl structure, and this cubic symmetry is reflected even in the smallest cluster size containing only four atoms. The abundance spectra [204] do not follow the shell-filling or jellium models and can be explained by assuming that the clusters grow by adding cubic subunits, whose local bonding configuration is identical to that of the bulk material. Even the smallest cluster can already be described as a tiny piece of the bulk lattice. This in contrast to what we see for metal clusters, where the structural units for small clusters are not equivalent to the bulk structure [163]. However, not all ionic materials behave as TiN, and a wide range of cluster structures has been reported for earth alkali oxides, and other ionically-bonded materials.

In addition to the electronic contributions, described above prior to our small excursion into TiN, the cluster stability is influenced by geometric and bonding contributions to their total energy, which have their origin in the local bonding configurations and interaction potentials. These geometric contributions become more important at larger cluster size when the electronic structure has already reached bulk values. A closer look at the different (geometric) energy contributions illustrates the modulation of shapes as a function of size for metallic clusters with nondirectional bonding and a 3D structure. The energy considerations are presented in detail in a recent review by Doye *et al.* [205], and we summarize here the salient points.

The cluster stability (or excess energy compared to the bulk material) is denoted as $\Delta(N)$ and is given by:

$$\Delta(N) = \frac{E_{\text{binding}}^{\text{cluster}}(N) - NE_{\text{cohesive}}^{\text{bulk}}}{N^{2/3}} \tag{4.2}$$

where $NE_{\text{cohesive}}^{\text{bulk}}$ is the bulk cohesive energy, and $E_{\text{binding}}^{\text{cluster}}(N)$ is the cluster binding energy normalized to the (estimated) number of surface atoms. The binding energy of the cluster is given as the sum of the contributions from volume atoms, which are entirely surrounded by bonding partners, and the contributions from surface atoms in different binding geometries, such as edge, kink, and terrace sites. With increasing cluster size the contributions from surface atoms becomes negligible, and the cluster stability asymptotically approaches the bulk value.

The binding energy of the cluster is the term which essentially determines the overall cluster shape and stability, and drives the transition from a spherical cluster to a faceted crystallite. The binding energy $E_{\text{binding}}^{\text{cluster}}$ is given by:

$$E_{\text{binding}}^{\text{cluster}} = h_{NN} E_{NN} + E_{\text{strain}} + E_{NNN} \tag{4.3}$$

where h_{NN} denotes the number of nearest neighbors and E_{NN} is the bond strength at optimum distance; the minimum of the interaction potential. E_{NNN} is the second nearest neighbor interaction, which in most cases is small enough to be neglected. The strain energy E_{strain} is reduced by the onset of faceting for larger cluster, and by local rearrangement of bonding configurations for smaller clusters as illustrated in Figure 4.7. Doye [205] derived a phase diagram extending from 10 to 80 atoms, which is presented in the review by Baletto *et al.* [206], and shows the rather dramatic modulation of structure and faceting as a function of cluster size using the Morse potential to describe bonding. The interplay between the different energy contributions leads to a considerable variability in geometric structures, which again defies the simple picture of geometric cluster stability illustrated in Figure 4.6, where layers of atoms are arranged in a progressive manner around a central atom.

4.4 Closing Remarks

Ultrasmall clusters are remarkable materials and occupy the transition regime between the single atom, molecule, and inorganic nanostructure. The richness of electronic and geometric structures and the rapid variation of properties as a function of size is unique. It is both an advantage and disadvantage in their application in functional nanostructures and devices, and an example of a metal cluster single electron transistor in Chapter 7.1 illustrates this point. On the other hand, metal clusters which already exhibit faceting can be synthesized with solution-based methods which stabilize facets not accessible in clusters grown by vacuum- or gas-phase-based approaches. From the point of view of industry, metal clusters or crystallites occupy a much larger production volume than all other nanomaterials described in this book. Catalysts are indeed some of the most important and ubiquitous materials, and any improvement in their activity, which emerges from fundamental studies based on our understanding of the material properties, is highly coveted.

5
Quantum Dots

A quantum dot (QD) is a zero-dimensional (0D) structure defined by confinement in all three directions of space. In that sense, the metal clusters discussed in the previous chapter can also be classified as quantum dots, however the label of "quantum dot" is most frequently used for semiconductor materials. In this chapter we will therefore focus on semiconductor quantum dots, their properties and synthesis. Quantum dots, like metal clusters, show the characteristic 0D density of states (DOS) (Chapter 1), which consists of sharp, well-defined energy levels akin to the electronic levels of single atoms and molecules [207, 208]. QDs are therefore often described as artificial atoms, and might be treated as such in many instances when describing electronic and optical properties. In analogy to atoms, it is possible to create quantum dot superlattices, and metamaterials, where the single dot is the basic building block [209–211] and the overlap of wavefunctions of adjacent dots takes on the role of the interatomic bonding.

5.1
Size and Shape in Quantum Dots

In Chapters 3 and 4 we discussed in detail semiconductor nanowires, where confinement is limited to two directions in space, and metal clusters with confinement in all three directions of space. The majority of examples focused on the electronic and geometric structure as a function of size. However, for QDs the shape [102, 212–215] is also highly variable and can be modified through the synthesis process. The name "dot" is therefore somewhat misleading, and for a better representation of these 0D systems we must go back to the original definition, which states that a QD is an entity where quantum confinement is present in all three directions of space. This means that a flat island, which is sufficiently small to provide

Inorganic Nanostructures: Properties and Characterization, First Edition. Reinke, P.

Published 2012 by WILEY-VCH Verlag GmbH & Co. KGaA

confinement along the direction of largest extension is also a QD and will exhibit the characteristic discrete DOS. For a complete description of QD electronic structure and optical properties, it is therefore necessary to take both parameters, size and shape, into account.

We will begin with a discussion of the influence of the size of QDs on the electronic and optical properties. We have already looked at the modification of band gap as a function of dimensions in some detail for semiconductor nanowires, and indeed the same relation holds true for QDs: the band gap increases with reduced QD diameter. However, we are now considering the transitions between discrete states, and the optical properties of QDs are defined by transitions between the sharp energy levels, which consequently often present as peaks with relatively narrow linewidth in optical spectra [216–220]. The measurement of absorption, emission, and luminescence spectra can therefore be used to directly probe the electronic structure of QDs. Although we only discuss optical spectra here in conjunction with QDs, they are highly sensitive probes in the characterization of semiconductors, and are without doubt some of the most versatile and important characterization techniques. The description of optical properties in the next section provides a short introduction into the optical properties of materials, and we refer the reader to the extensive library on optics and optical properties for a more in-depth study.

5.1.1
A Short Excursion to Optical Properties

An absorption spectrum is measured as follows: a sample with a thickness d is irradiated with light of intensity I_0 in the wavelength range of interest, and the intensity I of the transmitted beam is recorded. I is linked to the absorption coefficient α by the Lambert–Beer law $I = I_0 e^{-\alpha d}$ (it is assumed that reflection at the surface is negligible). If the wavelength of the incident light coincides with the energy of an electronic transition, the absorption coefficient will rapidly increase, a photon is absorbed, and the electron is excited to a higher energy level.

A wavelength (energy), which is larger (smaller) than the band gap of the QDs will not be absorbed and beam attenuation is minimal. However, once the photon energy is sufficient to trigger electron excitation across the band gap (HOMO–LUMO gap), a sharp rise in the absorption coefficient will be observed. Since the band gap of QDs is expected to increase with decreasing diameter, a shift of the absorption edge to smaller wavelength is expected. The band gap of the bulk material defines the absorption limit on the long wavelength side. The band gap of several bulk materials which are commonly used as QDs, are indicated in the graph included in Figure 5.1.

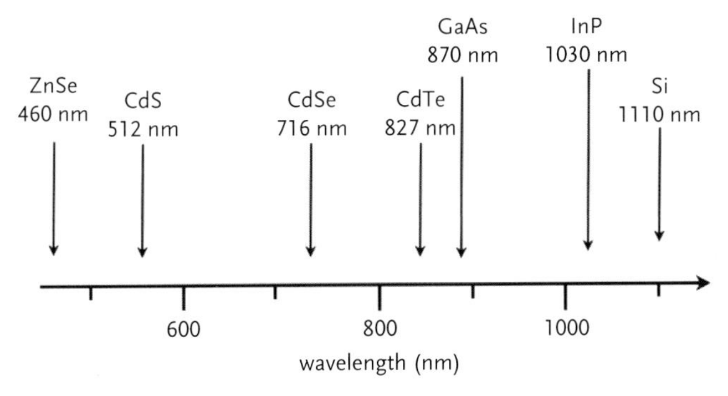

Figure 5.1 Band gap of some common semiconductor materials (room temperature). The band gap is the upper limit of the absorption edge when a photon has sufficient energy to excite an electron from the valence to the conduction band.

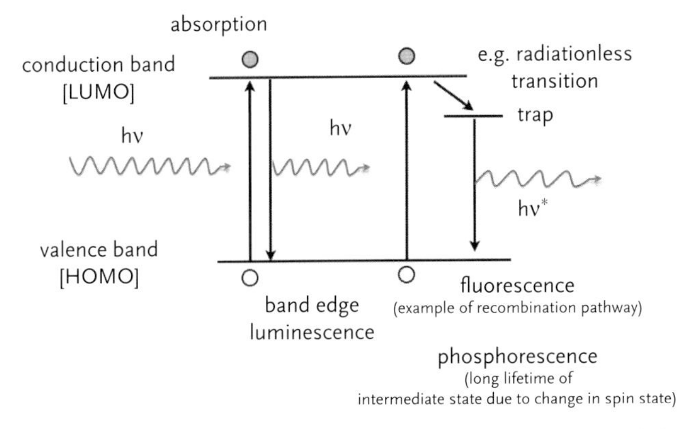

Figure 5.2 Schematic illustration of several optical processes, including absorption, and several pathways for recombination.

Once the electron occupies an excited state it can recombine with the hole in the valence band in several ways and two examples for recombination pathways are sketched in Figure 5.2. Firstly, direct electron–hole recombination (luminescence) with the emission of a photon whose energy is equal to the band gap, which is also the energy difference between ground and excited state. In bulk semiconductors the electron has a high probability of thermalizing to the conduction band edge, and recombination at the band-edge energy then dominates. Thermalization depends on the ability of the electron to lose small amounts of energy through interaction with other electrons or the lattice. All excitation and recombination processes have to obey energy and momentum conservation, and a direct transition, which does not require an additional particle, like a phonon, to guarantee momentum conservation, is faster and more likely than an indirect transition. Any transition which requires a change in spin state will also be slower than one where the spin state is not modified. The presence of very slow de-excitation processes can lead to the appearance of a long "afterglow" called phosphorescence. Secondly, the excited electron can recombine via states located in the band gap such as dopant levels, or defect states. It will then lose part of its energy during the de-excitation and if a photon is emitted in the final recombination with the hole, its energy will be smaller than the band-gap energy. The emission spectrum then contains information about the dopant levels, since they introduce states within the band gap. Photoluminescence spectroscopy, which measures the energy distribution of emitted photons after excitation with an intense monochromatic light source such as a laser, is probably the most widely used technique for the analysis of electronic structure and defect states of semiconductors and semiconductor nanostructures. Our summary of optical properties gives a glimpse of the complexity of optical transition processes, and should be sufficient to follow several examples included in this chapter.

5.2 Band Gap, Size, and Absorption Edge

Figure 5.3 shows the absorption spectra [222] of CdSe QDs as a function of size for QDs made by wet chemistry with the colloid method pioneered by LaMer [221] (see Section 5.3 for details on synthesis). The size range for the QDs is 1.2–11.5 nm, and a narrow size distribution could be achieved by repeated size-selective precipitation steps. The absorption edge can be easily identified, and it shifts from about 420 nm for the smallest QD to 650 nm for the largest QD, which is very close to the bulk value. The color of the QDs can be adjusted by size selection, and covers a large portion of the visible spectrum.

A reasonable description of the size dependence of the band gap for spherical QDs was developed by Brus [223] and is based on the effective mass approximation, and essentially treats the QD as a small chunk of bulk material, and consequently the dielectric constant of the bulk is used to calculate the energy of the first excited state. This approximation correctly reflects the trend of increasing band gap with shrinking size, but the gap energy decreases more rapidly than observed experimentally, especially for small QDs. A much improved theoretical description of the relation between CdSe QD size and band gap was achieved by Wang and Zunger [224] with a plane-wave pseudopotential method, and their study highlights several critical issues in the calculation of electronic states in QDs. A QD has a much larger percentage of surface atoms than a nanowire with the same diameter around its circumference, and contributions from surface atoms therefore must be

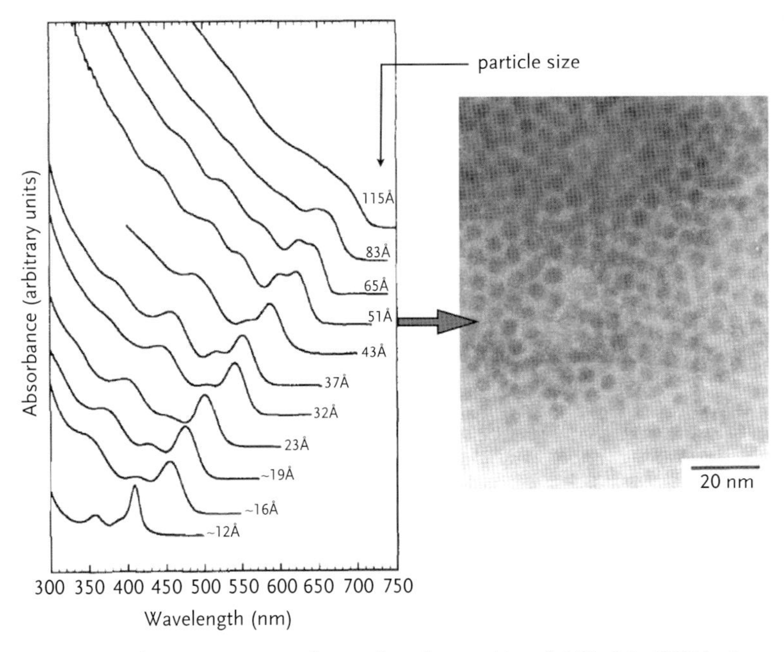

Figure 5.3 Absorption spectra of size-selected ensembles of CdSe QDs [222] in the size range between 1.2 nm and 11.5 nm. The transmission electron microscopy (TEM) image on the right hand side shows a densely packed array of QDs with an average size of 5.1 nm. Reprinted with permission from [222]. Copyright 1993 American Chemical Society.

included in any successful theoretical description of QD band gaps. For example, the surface in a compound semiconductor is often terminated by one element, which slightly offsets the stoichiometry, and induces a dipole and thus requires an additional electrostatic contribution to the QD energy. Surface strain, trap states, and shape distortions are additional challenges in the theoretical descriptions.

Two of many applications where the size-dependent absorption and emission characteristics of QDs are used in a very successful manner, are solar cells [218, 219], and selective tagging of cells in biology [225], where the QD serves as a fluorescent marker. The operation of a solar cell is a complex process; in simplified form it relies on (1) electron excitation via a photon and creation of an exciton, (2) charge separation of electron and hole, (3) charge transfer and (4) transport to the electrodes. A high yield solar cell is created by optimizing all of these processes simultaneously, which has led to the fabrication of structurally and compositionally quite complex solar cell assemblies.

QDs in solar cells have several advantages, which are discussed in detail in a recent review by Kamat [218]. The most obvious advantage is the ability to tune the absorption spectrum of the QDs by adjusting their size distribution. A CdSe QD ensemble can therefore be designed to absorb light across nearly the entire visible spectrum, and the solar cell can use a much larger fraction of light for charge creation (see chapter 7.3). An example for the use of QDs in biology are ZnS QDs, which are excellent fluorescent labels [225] with very little photo-bleaching and therefore superb long-term signal stability. Initially the very low solubility of the QDs in an aqueous solution was a major obstacle to their introduction in a "living system". This obstacle could be overcome by wrapping the QDs in micelles with the hydrophobic end oriented toward the QD. The micelles self-assembled around the QD, which was subsequently extracted, functionalized, and used in a ground-breaking experiment to mark single cells in frog embryos [167].

5.3 Synthesis of QDs

The synthesis of QDs meets the same challenges we have already encountered for other nanoscale materials: how can we control size, shape, and position? The main techniques in QD synthesis are based on organic and metal–organic synthesis, which delivers large yields of a wide range of QD materials, and Stranski–Krastanov (SK) type growth, where the QD nucleation is triggered by the relief of strain energy, which builds up at the interface between two lattice-mismatched semiconductor materials. The QDs grown by strain relief are of high crystal quality, immobilized on a substrate, and their position can be controlled by templated self-assembly. These two methods of QD synthesis are fundamentally different techniques, and are both used extensively.

5.3.1 QD Synthesis by Chemical Methods

The synthesis of QDs in solution is mainly used for ionic II–VI materials, although it has recently been extended to group IV and III–V semiconductors [209, 221, 226, 227]. Similar approaches have been used for the synthesis of metal clusters and the micelle-based synthesis of metal clusters, described in Chapter 3, has also been applied to semiconductor QDs. The synthesis of colloidal QDs, which produces QDs suspended in a solvent, begins with the addition of cation and anion precursors to an organic solvent containing a surfactant, which subsequently serves to stabilize the QDs. The precursors are typically metal–organic reagents, such as an amine or phosphine. A disadvantage of many metal–organic reagents is their toxicity and there is a considerable interest in finding alternative precursors with similar reactivity. CdO, for example, has been shown to be an excellent substitute for the highly toxic Cd-phosphines which were used in the early work on CdSe synthesis [228, 229].

The surfactant molecule, which is often a long-chain alkyl group with one polar group attaching preferentially to the ionic surface of the QDs. The surfactant attachment to the QD surface can be described as a dynamic solvation: the average coverage of the QD with surfactant molecules is stable, but they are constantly exchanged with the surrounding liquid [230]. The attachment–detachment kinetics determine the overall coverage of a given QD facet. The attachment of the surfactant molecule is therefore weak enough to allow for the QD material precursors to continue the QD growth, but sufficiently strong to block intra-QD interaction and protect them against agglomeration (see Figure 4.3). The surfactant therefore is a critical element in the control of QD growth and, as we will see later, can also be used to control QD shape.

A frequently used surfactant compound is *n*-trioctylphosphine (TOP), where three octyl groups with the formula C_8H_{17} are attached to a central P atom, or *n*-tri-octylphosphine-oxide (TOPO), where an oxygen group is attached to the phosphine end:[1]

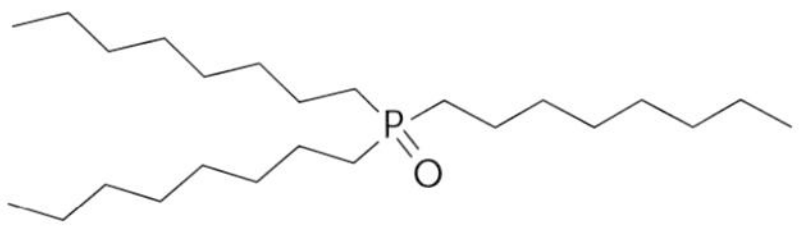

It is also possible to substitute O by S, or Se, which makes the TOP group the carrier of an anion, which can function as a coordinating group at the QD surface.

The interaction of the surfactant with the surface of the QD is one part of the reaction cycle, the other one is the dissociation of the anion and cation precursors to form the QD material. The anion/cation precursors dissolve in the solvent and after a sufficient degree of supersaturation has been reached, they will react to form nuclei of the ionic solid, which will then grow. The availability of cation/anions to the growth of the QD is controlled by the precursor concentration and the strength of bonding of cation/anion in the precursor state. The strength of bonding between organic group and cation/anion in the metal–organic precursors determines the ease of reaction between cation and anion, and thus the reaction rate for the formation of the QD [227]. If the bonding is too strong, the crystallites will be small and high reaction temperatures are required; if the bonding is weak, large crystallites will form very rapidly and control over the size distribution cannot be achieved. The capping agent or surfactant at the same time serves to keep the QDs in solution, and stabilizes

[1] From: http://www.chemicalbook.com/

the colloid. The type of cation or anion can be changed during the reaction cycle allowing for the formation of core-shell structures [231, 232]. Unfortunately simply mixing the reactants will not deliver the desired narrow size distributions.

A narrow size distribution can only be obtained if the nucleation process is a temporally discrete event, and all nuclei form in a very narrow time interval. The temporal discreteness can be achieved if one or both precursors are injected into the solvent to instantly create a high level of supersaturation, which triggers the nucleation process [209, 233]. The overall process of nucleation and growth is illustrated schematically in Figure 5.4. The temperature drop which often occurs as a consequence of the reactant injection also contributes to the abruptness of the nucleation event. The supply of reactants and the temperature of the solvent are then used to control the growth of the nuclei, and subsequent Ostwald ripening. Aliquots of the solution can be removed at different points of the growth process to obtain QD samples of all size ranges. The QDs are capped, and therefore protected against agglomeration (see Figure 4.3). A typical reaction sequence with TOP as an organic solvent will lead to a TOP-terminated

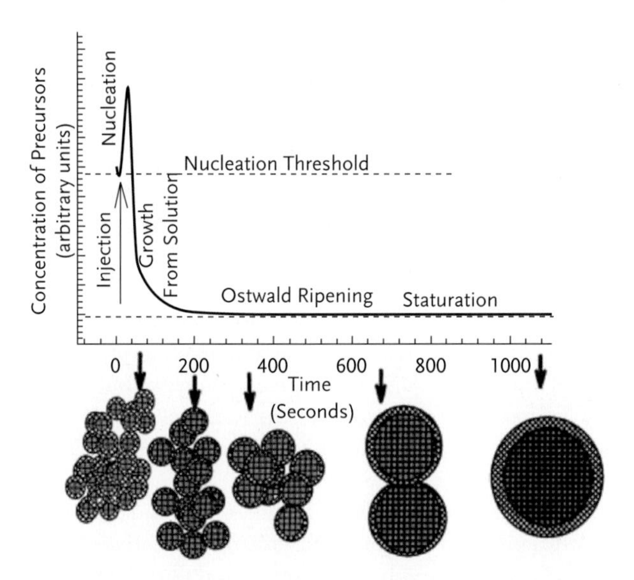

Figure 5.4 Schematic illustration of the size-selective growth [209] of colloidal QDs. The starting point of the reactions is the injection of the metal–organic precursor into the solvent (marked in graph). The large supersaturation, which is created by rapid injection leads to a near instantaneous nucleation. The nuclei possess a narrow size distribution, which then undergo an Ostwald ripening process. The samples are then removed from solution at certain time intervals, which define the average QD size. Reproduced with permission of Annual Review of Materials Science copyright 2000 by Annual Reviews Inc. from [209]; permission conveyed through Copyright Clearance Center, Inc.

QD surface. The capping molecules can be functionalized or exchanged for other groups by subsequent chemical treatment.

Reiss *et al.* [233] predicted and it was later shown experimentally that a tight control of supersaturation can lead to a sharpening or "focusing" of the size distribution [230, 234]. Focusing, which is a substantial narrowing of the size distribution, is an excellent example for the application of nucleation theory, and at the same time illustrates the intricate interaction between the solid and the liquid phase in solution based QD synthesis. In Figure 5.5 the growth rate and critical size of nuclei is illustrated as a function of time for different supersaturation of the precursor; the equivalent in the gas phase is a variation of the reactant pressure. A negative net growth rate is observed if the nuclei size is smaller than the critical nucleus and there is a high probability of dissolution. Once a nucleus reaches the critical radius, the growth rate crosses to positive values and we move from the nucleation to the growth stage. If the supersaturation is kept constant, the nucleation will be followed by a short period of very rapid growth with concurrent increase of the nanocrystal radius. The growth rate reaches a maximum and will then slow down due to a decreasing surface–volume ratio, and, from a purely geometric argument, now a larger number of atoms is required to achieve the same gain in radius.

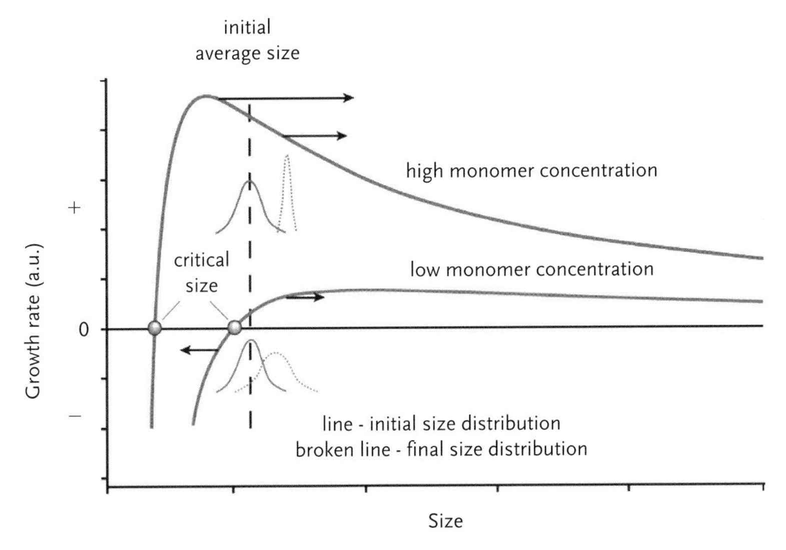

Figure 5.5 Illustration of size-distribution focusing, adapted from Yin and Alivisatos [230]. The graph shows the growth rate as a function of QD size for high and low monomer concentrations. The critical size is indicated for both cases, and any crystallites whose size is below this limit, will likely dissolve, while those above this size limit will grow. The average size is chosen to lie just above the critical size for low monomer concentrations, and is thus for the high monomer concentration relatively far above the critical size and none of the crystallites is sufficiently small to dissolve. Reprinted by permission from Macmillan Publishers Ltd: Nature [230], copyright 2005.

After a rapid nucleation process the average size of the distribution moves to larger radii and crosses the critical radius. A distribution where part of the distribution (the smaller crystallites) is still on the left-hand side of the critical radius will experience a loss and dissolution of the small crystallites, while the larger ones continue to grow and the overall distribution therefore broadens. This is evident in Figure 5.5 for the case of low precursor concentration. A focusing (narrowing) of the distribution is achieved by injection of a large amount of precursor molecules to move the whole distribution of sizes to the right-hand side of the maximum growth rate, which is the slow growth side of the distribution. To be precise: the distribution itself does not move, but its position with respect to the growth rate–size curve is changed if the supersaturation is modified. The smaller crystallites now grow faster than the larger ones, and will no longer dissolve: Ostwald ripening is suppressed and the overall distribution narrows.

The QD size and size distribution is the primary critical aspect in QD synthesis since it defines the extension of the confinement potential and thus whether a nanocrystal will indeed behave like a 0D system. However, manipulation of the QD shape allows a change in the relative percentage of surface atoms, control of the surface structure and contributions from different facets, which can influence overall reactivity, and promote shape-specific optical properties such light polarization or plasmon propagation [212, 215, 228, 235–241].

The shape of a crystal is determined by the relative growth rates of the facets: the fast growing facets disappear as the size of the crystal increases, and only the slow growing facets remain and thus determine the overall shape [242]. The fastest growing facets are typically the ones with the largest surface energy and thus the largest chemical potential gradient as a driving force for growth and adsorption of ad-atoms from solution. The equilibrium shape is consequently the one with the lowest overall surface energy. Modulation of shape can therefore be achieved if the surface of the crystal or the interfacial energy between crystal surface and liquid is changed, thus promoting or depressing the growth of selected surfaces. While this might sound like a straightforward assignment, it is difficult to modulate surface and interfacial energies in a controlled manner, mostly due to the challenges in measuring surface energies for select crystal planes, and the complexity of the solid–liquid interface. However, the empirical knowledge and understanding of shape modulation for QDs grown from the liquid phase is already substantial and many shapes can nowadays be fabricated at will.

An excellent example for shape modulation is the evolution of CdSe and CdTe nanocrystals in the QD size regime: the shapes obtained by selective solution-based processing include spherical, cylindrical, branched, rice-grain or spindle, arrow, and tetrapods. Figure 5.6 shows transmission electron microscope images of these nanocrystals [212, 241, 243]. The thermodynamic equilibrium shape is a near spherical nanocrystal with low surface energy facets and a favorable surface–volume ratio, and it is achieved if the growth

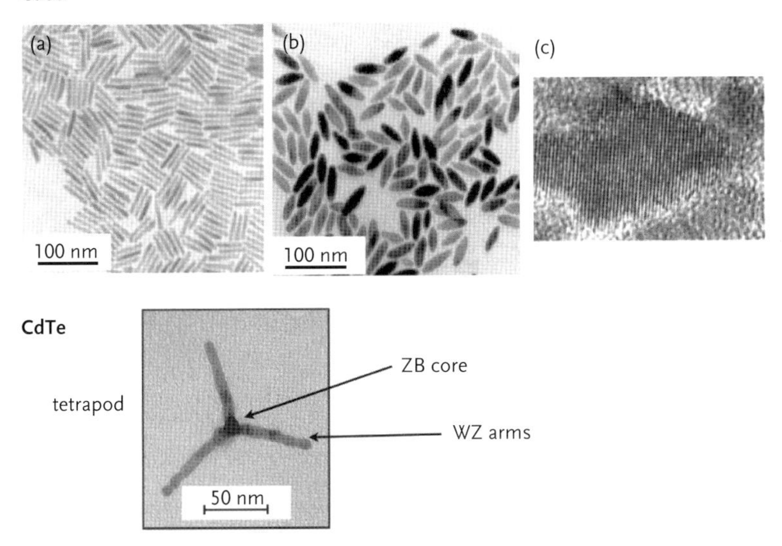

Figure 5.6 Examples and electron microscopy images showing different shapes which can be obtained for CdSe and CdTe colloidal QDs, (a) cylinder [228], (b) rice-grain, (c) high resolution image of an arrow-shaped crystallite [243]. The bottom image is a CdTe tetrapod [238], the image is a projection onto the imaging plane and the fourth arm is normal to the plane of view. Reproduced with permission from Advanced Materials [228]. Copyright Wiley-VCH Verlag GmbH & Co. KGaA 2003. Reprinted with permission from J. Am. Chem. Soc. [243]. Copyright 2000 American Chemical Society. Reprinted by permission from Macmillan Publishers Ltd: Nature Materials [238], copyright 2003.

rates are relatively slow and one operates in the regime of low supersaturation.[2] The overall shape variation as presented in Figure 5.6 results from modification of the surfactant and precursor concentrations, which promote or inhibit the growth of certain facets selectively. A detailed discussion of facet blocking and growth is given in the review by Scher *et al.* [241].

The CdTe tetrapods present a particularly intriguing shape with a considerable complexity: four crystalline "arms" with wurtzite (WZ) structure extend from a central core with zincblende (ZB) structure. The WZ arms are attached at the $\{111\}$ facets of the ZB core, which is equivalent to the $(000\bar{1})$ plane in WZ. This common facet is then the starting point for the growth of the WZ arms, and also terminates the individual arms. The crystallography of the WZ and ZB structures is discussed in detail in Chapter 3 on semiconductor nanowires. The energy difference between ZB and WZ polytypes is sufficiently small to allow for the emergence of polytypism in nanocrystals, and sufficiently large to make selective

[2]Supersaturation is defined by the ratio of the present concentration of the reactant to the equilibrium concentration of the reactant.

nucleation and growth of one phase possible. The initial CdTe nucleation occurs in the ZB structure but the growth is pushed toward the WZ structure by using a surfactant mixture, which stabilizes the nonpolar side facets of the WZ arms. The tetrapod synthesis is therefore achieved through a combination of selective phase nucleation in a system with polytypism, and the targeted action of a surfactant. The length and thickness of the WZ arms can be modified by changing the Cd/Te precursor ratio, and by using a mixture of two different surfactants. The diameter of the WZ arms is typically a few nanometers, and thus still in the confinement regime. The modulation of the WZ arm diameter has been used to modify the optical spectra of the tetrapods [215, 235, 236, 243].

Shape transformation in the synthesis of CdSe nanocrystals, which is shown in Figure 5.6, progresses from sphere to rice-grain shaped to cylindrical, rod-like with increasing surfactant monomer concentration. The shape modification can be explained by selective adjustment of the surface energies and thus the growth rate of the different crystallite facets and surfaces. The facet-selective change in surface energy can be achieved by (i) strong bonding of the surfactant and thus blocking of reaction sites for growth; (ii) preferential adsorption of one type of surfactant molecule; the concentration ratio of two surfactants can then be used for selective modulation of facet growth rates; (iii) choice of a surfactant with facet selective bonding strength – for example, hexylphosphonic acid (HPA) binds more strongly to cation-terminated surfaces. Let us assume that only one facet is weakly blocked by surfactant molecules: it will grow exceptionally fast and a rod-like shape will form; the arrow shapes are presumably formed if the facet growth is so rapid that layers are not completed before the next layer nucleates, building a wedding-cake like structure, and rice-grain shapes are intermediates between spherical and rod shapes. It can be seen that the surfactant-modulated variation of surface energies and growth rates of individual facets leads to a wide range of shapes, which can be adjusted at will.

Shape control via the modulation of relative surface energies is also applied to the synthesis of metal nanocrystals [236, 237, 240, 244], and parallels the synthetic approaches used for the semiconductor materials. The interest in shape control for metal nanocrystals is mainly driven by optical applications related to plasmonics and surface enhanced Raman scattering (SERS). The size of the nanocrystals is usually above 10 nm, and they can therefore no longer be considered as 0D systems in terms of their electronic structure. However, many of their optical properties are modified and often enhanced by choice of a specific shape in the size regime of a few hundred nanometers. The optical response for SERS and surface plasmon resonances (SPR) is very sensitive to only small variations at the surface. The attachment of even small amounts of adsorbates can trigger a characteristic shift in the SPR signal, which makes the metal nanoclusters excellent sensors.

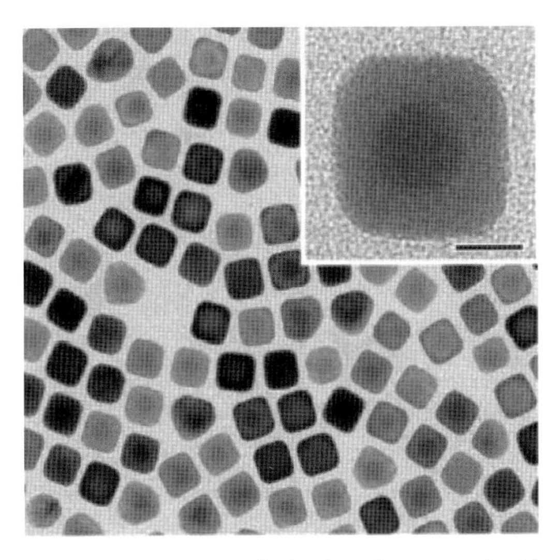

Figure 5.7 TEM image of cube-shaped nanocrystals with an average diameter of 20 nm. The core is a Au seed crystal and the shell is a Ag cube [245]. The cube quality, Ag and Au layer thickness, and optical properties were studied in detail as a function of the processing parameter such as precursor concentrations, surfactant type and concentration. Reprinted with permission from ACS Nano [245]. Copyright 2010 American Chemical Society.

One particularly intriguing example is the synthesis of Au and Ag cubes, which was achieved in 2002, and an example for the Au–Ag core shell cubes [245] is shown in Figure 5.7. The crystal habit for an f.c.c. metal is determined by the ratio of {111} to {100} facets, where the dominance of {111} leads to an octahedral shape, for {100} facets we will obtain a cube. If the corners of a cube are truncated {111} facets are exposed, and truncation of the edges exposes {110} facets. Therefore, if the reaction conditions are adjusted to promote growth in the <111> direction, the cubes are formed, and the {111} facets disappear. The relative growth rate of the facets is controlled by the interplay of temperature, precursor concentration, and the surfactant type and concentration. The array of shapes, which have been synthesized in metal and semiconductor nanocrystals and QDs is vast, and now offers the unprecedented possibility to tailor size as well as shape.

5.3.2
Strain-Driven Self-Assembly – Stranski–Krastanov Growth

The strain-driven synthesis follows the Stranski–Krastanov (SK) growth mode, and it offers superb control of QD size distributions, and yields "naked" QDs without surfactant coating which are immobilized on a substrate. Control of their spatial distribution has been achieved by patterning the substrate to create preferential nucleation sites [246], or through strain engineering, which is used in the fabrication of multilayer QD stacks and superlattices [247].

Strain-driven synthesis of QDs is a form of molecular beam epitaxy (MBE), where a clean substrate is exposed to a well-defined flux of atoms or molecules, which then condense on the surface to form a thin film whose morphology is controlled by strain, temperature, and reactant flux. MBE is a highly successful method for the growth of complex structures such as semiconductor quantum well superlattices, and it was noticed in the late 1980s that quantum well growth with dissimilar materials (heteroepitaxy) can lead to dramatic changes in the overlayer morphology and favors the formation of small crystalline islands which function as QDs. It was soon recognized that the appearance of QDs is linked to the build-up of strain energy in the overlayer due to lattice mismatch with the substrate.

Figure 5.8 summarizes lattice constants and band gap for several common semiconductor materials (cubic lattice). A mismatch in lattice constant drives the nucleation of island morphologies, and the mismatch in band gap defines the confinement of electron and holes on the island, which then becomes a QD. A selection of well-known material combinations includes (QD/substrate) Ge/Si, InAs/GaAs, GaSb/GaAs, InP/GaInP, InAlAs/GaAs, InAs/InP [248–256]. The strain of a QD system can also be modulated in a controlled manner by using alloys, such as Ge–Si mixtures for the growth of QDs on Si(100) [257], and InGaAs instead of InAs on

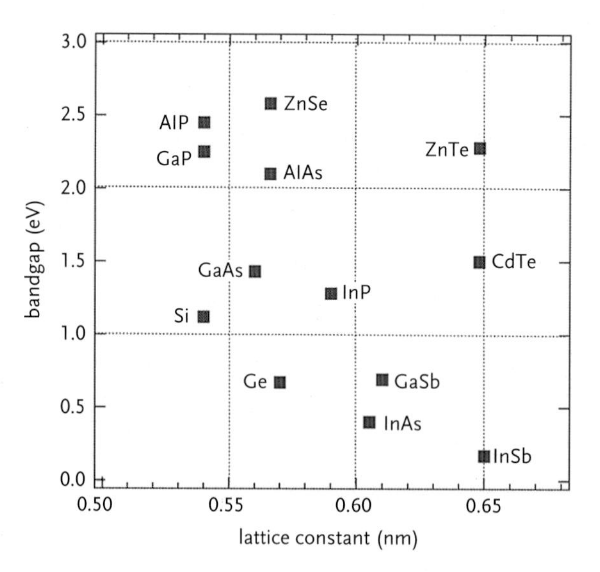

Figure 5.8 Summary of band gap and lattice constants for semiconductor materials commonly used in strain-driven Stranski–Krastanov growth of QDs.

GaAs [220, 258, 259]. It is possible to form islands by SK growth in metallic systems [260] such as Ni on Ru(0001). The upper limit for quantum confinement has been discussed extensively throughout this book, but there is also a lower size limit for confinement in QDs: if the QD becomes too small, the ground state energy level for a charge might move to an energy which lies above the energy of the confinement potential. In this case, the QD can no longer hold a charge and simply functions as a shallow potential barrier.

Typical SK growth in a lattice-mismatched semiconductor system progresses as following [261, 262] (see Figure 5.9): the growth starts with the formation of a so-called wetting layer, where a uniform layer (UL) builds up to a critical thickness, which is specific to the material combination of substrate and overlayer. The nucleation of coherent islands or QDs with relatively narrow size distributions occurs once this critical thickness has been surpassed. A coherent island (CI) has an epitaxial relation to the underlying substrate and does not contain any dislocations. If the deposition process is continued, these islands will grow in size and frequently undergo a series of material specific changes in island shape, until dislocations appear, and the island structure switches from coherent to dislocated islands. The sequence of transformations from wetting layer to dislocated island is discussed in chapter 5.3.3 for the Ge–Si system.

The strain energy is clearly a critical parameter in the SK growth of QDs, and has its origin in the lattice mismatch between substrate and overlayer. Strain in the lattice is caused by a force acting on the lattice and leading to displacement of the individual atoms. This can be envisaged by thinking of the lattice as atoms connected by springs, which represent interatomic bonding: if we now compress or elongate the springs by exerting a force, the displacement of the atoms with respect to each other will be controlled by the force–distance relation defined by the interatomic potential energy curve (see Chapter 2). The change of atom position within a lattice in response to the applied force is described by the displacement vector $\vec{u}$. Stress σ is

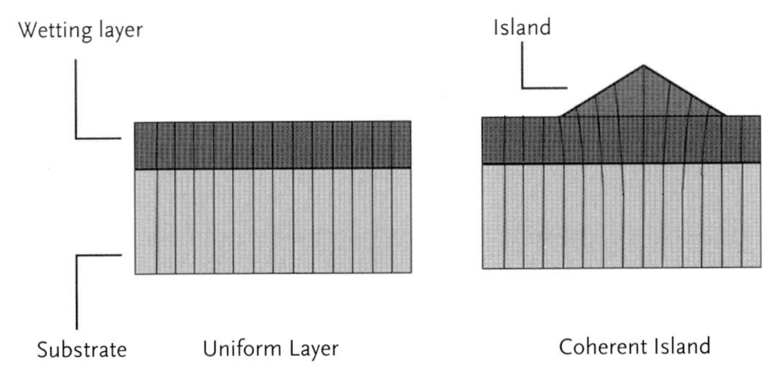

Figure 5.9 Illustration of the lattice continuity as it is observed in layers and islands during strained layer epitaxy in the Stranski–Krastanov growth mode.

defined as F/A (force per unit area), and the reaction of the lattice, the displacement of lattice atoms in response to the application of a force, is the strain ε. Both stress and strain are tensors within the coordinate system of the solid, and are connected by Hooke's law if the deformation is entirely elastic. An in-depth discussion of stress and strain components for quantum wells and dots is given in Refs [207, 263], and a more general discussion of the elastic deformation of solids is included in many advanced materials science textbooks. In a planar system, such as a wetting layer on a substrate, lattice mismatch, which determines the in-plane strain is given by:

$$\varepsilon_{in\text{-}plane} = \frac{a_0 - a_L}{a_L} \tag{5.1}$$

where a_0 is the lattice constant of the substrate, and a_L is the lattice constant of the overlayer/wetting layer. The stress is in plane since it is due to the lattice mismatch between the two materials, but the strain has not only a lateral component but also a component perpendicular to the interface due to the Poisson relation. The coupling between different directions is described by a matrix of elastic stiffness constants, which is a tensor whose components can be derived from the lattice geometry.

The strain energy in the wetting layer (uniform layer, UL) increases with layer thickness, and the elastic energy per unit area for a UL is given by:

$$\gamma_{elast} = t\,\varepsilon \frac{Y}{2(1 - \nu^2)} \tag{5.2}$$

where t is the film thickness, Y is Youngs modulus, and ν is the Poisson number. A derivation of this expression can be found in Refs [263] and [264]. The total free energy of the system is composed of the strain energy, and all interfacial and surface energies.

The formation of a QD or a coherent island can only occur if the system moves into an energetically more favorable state: the energetic cost to making the surfaces of the QD, and the interface between QD and substrate must be balanced by the reduction in strain energy due to lattice mismatch. The comparison is made between the energetic cost of making a QD and the further increase of the wetting layer thickness. These contributions to the energy of the system depend sensitively on the kind of facet on the QD and thus surface energy, the size of the QD, and the thickness of the wetting layer. The transitions between uniform layers, coherent islands and dislocated islands have been described successfully in several publications based purely on thermodynamic considerations [261, 262, 265–267], although the discussion about the relevance of kinetic contributions to the transformation of QD shapes during the growth cycle is ongoing.

The size distributions of QDs formed by SK growth are rather narrow, and appear not to exhibit Ostwald ripening, which would lead to a broadening of the distributions. Following the thermodynamic argument developed above, the nucleation of coherent islands presents a thermodynamically stable configuration and should therefore lack a driving force for further ripening of the islands. A counter argument can be made, that the coherent islands might not present the thermodynamically stable configuration, but that ripening is kinetically hindered, for example, by a kinetic barrier to facet growth, and size distributions therefore remain largely stable [268]. Observation of a narrow size distribution is therefore not a unique signature of a thermodynamically stable QD phase. Shchukin and Bimberg [265] lay out many of the arguments in this discussion in their review on ordered nanostructures, which includes faceting and QD formation. Spencer *et al.* [269, 270] discuss the morphological instability of strained uniform films, which includes the formation of coherently strained islands, and indicate that low temperatures during the growth process and high deposition rates impose kinetic limits. Gray *et al.* [271, 272] exploited this regime, of kinetically driven growth with high deposition rates and relatively low temperatures described by Spencer *et al.*, in the fabrication of so-called QD molecules, which arrange around pits in the wetting layer. The possibility of moving from processing conditions where the growth can be either thermodynamically or kinetically controlled leads to a complex multitude of QD-related structures.

The shape of QDs is determined by the complex interplay of strain, surface and interfacial energies, and a wide range of shapes has been reported, including pyramids, square and rectangular huts, ellipsoids, truncated pyramids, and probably many more. QD shape is not only specific to the material system but also changes during the growth process [261, 273]. The shape has a profound influence on the electronic and optical properties of the QDs, firstly, because it directly determines the geometry of the confinement potential, and secondly, because the shape is coupled to the strain field within the dot [274]. Grundman *et al.* [251] showed the interplay between strain field and electronic properties for pyramidally shaped QDs. The strain is largest at the base of the QD where it is connected to the wetting layer, and smallest in the center of the dot. The local distortion of the lattice modifies the electron and hole wavefunctions and thus the local effective mass of the charge carriers (assuming that effective mass is still a valid description for small systems). As a consequence the energies of the electronic states are modified, which will be, for example, visible in a shift of the peaks in the photoluminescence spectra.

An example in this context is the application of so-called strain reducing layers (SRL) to shift the main emission line in InAs (on GaAs) QDs to 1.3 μm, which is highly coveted to couple QD lasers to fiber optic communication systems [220, 258, 259]. The SRL is a GaInAs layer, which is used to cap the InAs QD array, and shifts the InAs QD emission line in the desired direction compared to a GaAs capping layer. The shift of the InAs emission line is proportional to the In content in the SRL, and has thus been correlated to the amount of strain between QD and capping layer. The use of GaInAs, however, also modifies the characteristics of the band alignment and confinement potential at the QD–SRL interface, but the strain modification appears to be the dominant contribution to the emission line shift.

5.3.3 The Ge–Si System – Shape Evolution During Growth

As a last step we will now observe the evolution of QD shape as a function of nominal overlayer thickness[3] for Ge QDs on Si(100) [214, 268, 273–276]. The transition from wetting layer to dislocated islands is particularly well documented for the Ge–Si system, and following the entire sequence provides excellent insight into the general principles governing QD growth. The Ge–QD growth is summarized in Figure 5.10. In a Ge/Si system the typical wetting layer thickness is 0.1–0.15 nm, followed by the formation of coherent, pyramid-shaped islands bounded by {105} facets, which switch to domes with a more complex set of facets at about 4.5 nm, and dislocated domes are formed at a nominal layer thickness of about 10 nm.

The wetting layer exhibits the typical reconstruction of a Ge(100) surface and includes dimer vacancy lines (DVLs), which appear as extended defect structures in the STM images. The DVLs are a mechanism of strain relief within the wetting layer. Once the critical thickness is reached, hut clusters nucleate and populate the surface. These hut clusters are bounded by four {105} facets, which have an exceptionally low surface energy due to a dimer reconstruction at the surface. The low surface energy favors the formation of the {105} facet over other low index facets, which are observed in many other QD systems. The next step is a ripening of the symmetric hut islands with a square base to form elongated hut islands, which is a kinetically-controlled modulation of the shape. Detailed analysis of the transformation to rectangular hut islands shows that the nucleation barrier for the growth of a new layer on a {105} facet is the kinetic bottleneck in island growth. The facets grow line-by-line from widest point, the bottom of the facet, to the top, which is the narrowest point.

The nucleation barrier of a new facet layer increases with facet size, and decreases with facet filling: the more of a facet is filled, the smaller the barrier to grow the next line. The addition of a new facet layer to a large island is therefore more difficult than for a small island, which narrows the overall size distribution. The preference for nucleation on a small facet also perpetuates the elongation of originally square islands. The origin of

[3] The thickness of an overlayer/thin film is only properly defined if we discuss flat layers such as the wetting layer. However, as soon as islands begin to emerge, the concept of a layer thickness breaks down. We therefore use a "nominal" thickness, which is the thickness of a hypothetical, flat layer, which contains all the material in the islands. The number of atoms per monolayer is approximately 5×10^{14} atoms cm^{-2}.

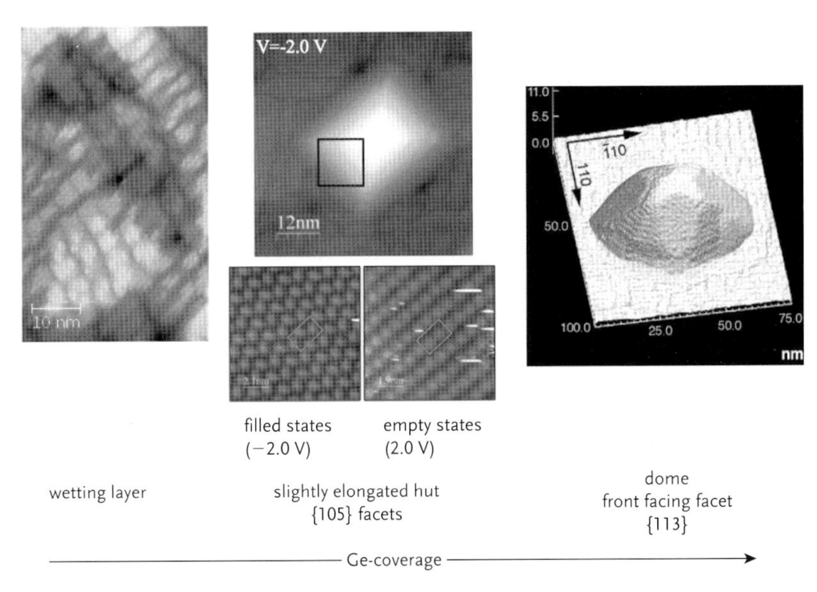

Figure 5.10 Shape evolution of Ge–QDs during the SK growth with increasing Ge coverage. The STM images show the wetting layer (Ge(100) reconstructed surface with dimer vacancy lines) and hut clusters [$T = 450\,^\circ$C, 0.0032 nm s^{-1}, 2 ML (WL), and 4 ML (hut) of Ge] including an empty and a filled state image of the surface reconstruction on the {105} facet (courtesy of C. A. Nolph). The image on the right-hand side [273] shows a dome obtained for growth at 600 °C for a nominal Ge coverage of 10 ML. The image of the dome is from Science [273]. Reprinted with permission from AAAS.

the observed dependence of the activation barrier on facet size is not entirely clear, and it might depend on the strain field distribution within the QD and at the transition to the wetting layer. Once the hut clusters are large enough they transition to dome-shaped clusters with {113} and several other types of facets. These facets have a steeper angle with respect to the plane of the wetting layer and therefore provide a more efficient strain relief. The formation of domes is then followed by the appearance of dislocated dome-shaped islands, which are rather large and behave electronically like bulk Ge.

5.4 Superlattices Made of QDs

In analogy to the formation of solids from atoms it is also possible to design superlattices made entirely of QDs, which can be realized with colloidal QDs as well as with QDs grown in the Stranski–Krastanov mode [210, 247, 277–282, 283, 284]. The interest in superlattices is motivated by the promise of novel properties, which emerge from the cooperative interaction between QDs within the superlattice, and the periodic structure of the superlattice. Superlattices can be fabricated in two and three dimensions, and the precision in their assembly with nanometer periodicity by far surpasses current capabilities in top-down fabrication with lithography methods. The QD[4] superlattices might be developed into so-called metamaterials, where the collective properties of a material are enhanced and can often be quite different from those of the individual building blocks. QD superlattices are relatively new materials, and their integration in applications is still in its early stages and will depend on the development of strategies which allow synthesis of high quality material where order extends over large areas and volumes.

Binary superlattices made of colloidal QDs have been envisaged as three-dimensional (3D) magnetic data-storage units, as sensors, and are often mentioned as the basis for the formation of photonic band-gap materials [210, 281, 282, 285]. The emergence of a photonic band gap can be understood in analogy to the formation of an electronic gap, which is the consequence of the Bragg condition, fulfilled by an electron wave confined within the lattice. The propagation of an electromagnetic wave and conditions for the formation of standing waves are controlled by periodic variations of the dielectric constant in a solid, which can be defined by the use of different materials within a superlattice. An example of a naturally-occurring photonic band-gap material and superlattice is opal, where colloidal silica spheres are arranged in a dense packing [210, 286]. However, the binary superlattices made from colloidal QDs show a large variability in structure, which makes them on the one hand more intriguing as objects for fundamental studies of crystallization, and on the other hand offer a potentially a wider spectrum of applications. However, while this is a highly coveted goal, a photonic band-gap material from colloidal QDs prepared by self-organization has not yet been presented.

The formation of long-range, spatially-ordered superlattices from colloidal QDs is an excellent model system to study fundamental aspects of ordering in a hard-sphere system [280, 287–290], and to investigate the impact of different types of interactions such as charges, van der Waals, and dipolar forces on the crystal structure [291]. The size and composition of colloidal QDs, which are used to build superlattices, can be modified independently, and sometimes bimodal size distributions of the same material are used. However, binary superlattices with two different materials for the QDs are much more common and interesting.

[4] In many publications which focus on synthesis, the nanometer-sized crystals are called quantum dots (QD), and in the majority of publications on superlattices the term nanocrystals (NC) is used, while actually discussing the same nanoscale material. In this book the term QD will be used throughout this chapter, and we only want to alert the reader here to the different conventions which are used in the literature.

The assembly of long-range, spatially-ordered superlattices from hard spheres has been investigated for quite some time, to study the thermodynamic and kinetic aspects which govern the emergence of densely packed lattices [286–288, 290]. A hard-sphere ensemble is characterized by a purely repulsive potential, which can be considered the equivalent to the repulsive component stemming from the Pauli exclusion principle on the atomic scale. In macroscopic spheres the repulsive term prevents merging of individual spheres. Prior to the assembly of QD superlattices, many experiments have been performed with polymer, latex, or silica spheres of different size ratios representing binary mixtures. However, the hard sphere model has been found in many instances to be equally suitable to describe the assembly of colloidal QD superlattices, albeit that a small attractive potential is always present due to the van der Waals interactions between the organic surfactant molecule shells surrounding the QDs. This attractive force can be modulated to some degree by the choice of solvent.

A densely packed binary superlattice made of A and B has a packing density close to or even slightly larger than an f.c.c. lattice, which has a volume fraction of 0.7405. Based on space filling arguments the stoichiometry of the densely packed structures can be AB, AB_2, and AB_{13}, which are isostructural to NaCl, AlB_2, and $NaZn_{13}$, respectively for a ratio of r_A/r_B between 0.5 and 0.8 where A is the larger sphere [282, 286, 288, 290–293]. Murray *et al.* [288] give a detailed derivation of the packing geometries and stoichiometries in binary solids, which lead to the formation of stable densely packed structures. All other stoichiometries even if they lie within the desired radius ratio will lead to structures, which are not densely packed, and phase separation, amorphicity, or random alloys, and even fivefold quasicrystal symmetry can be observed [294, 295]. The densely packed superlattices from hard spheres are thermodynamically stable and will form if the mobility of the constituents is sufficient. Synthesis begins with mixing a solution of the two QD materials and subsequent evaporation of the solvent. A dilute QD liquid is therefore transformed into a densely packed superlattice and the kinetics of solvent evaporation are closely connected to the kinetics and motion of the QDs during the "freezing" process.

The ordering into superlattices is driven by the entropy S contribution to the free energy F, which is given by $F = -TS$ since the internal energy term U goes to zero in a hard sphere system. The crystallization of the superlattice can therefore only be driven by entropy. The entropy term is composed of a configurational, and a free volume/translational entropy contribution, where the configurational entropy describes the degree of ordering in a given volume, while the translational entropy describes how far a sphere can move from its center of mass before it collides with its neighbor. The increase in translational entropy overcompensates the loss in configurational entropy and drives the crystallization.

The synthesis of densely packed superlattices from colloidal QDs was demonstrated by Redl *et al.* [210] using PbSe and Fe_2O_3 QDs for the assembly of AB_2 and AB_{13} hard-sphere superlattices, and by Shevchenko *et al.* [282], who expanded the range of superlattices beyond the hard-sphere systems and presented a large number of superlattices with a wide range of structures. Two representative superlattices are shown in Figure 5.11. However, the simplicity of the system is deceptive, and control over the crystallinity of a superlattice can be difficult to achieve, and competition between different crystalline and amorphous phases often leads to lack of more long-range order. One of the major challenges in the use of colloidal QDs is the width of the size distribution, and its impact on the stability of a structure [291, 296], and the subtle interplay between the organic surfactant coating of the nanoparticles and the solvent. A small difference in the attractive potential and thus deviation from the ideal hard sphere model, can therefore lead to a substantial modification of the overall interparticle interaction and thus favor a different structure. This aspect was illustrated by Shevchenko *et al.* [282], who showed the coexistence of 11 structures for a superlattice assembled from 6.2 nm PbSe and 3.0 nm Pd colloidal QDs. However, while control of the long-range structure in a superlattice can be difficult to achieve at times, the structural diversity is a great advantage, and gives access to a wide range of metamaterials made from many different building blocks.

In the strain-driven synthesis of QDs (Stranski–Krastanov), a superlattice can be formed by the sequential growth of multiple layers of QDs, which

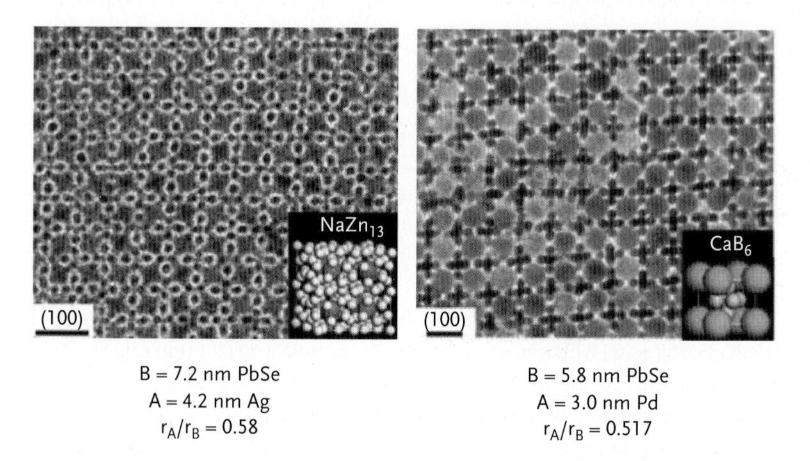

Figure 5.11 TEM images of superlattices built from colloidal QDs; the images show only a fraction of the examples given in the associated publication [282]. The $NaZn_{13}$ structure on the left hand side is an example of a densely packed structure. The inset shows the structure in the direction of the projection onto the imaging plane of the microscope. Reprinted by permission from Macmillan Publishers Ltd: Nature [282] copyright 2006.

are separated by a thin semiconductor spacer layer. The main goal for these type of QD superlattices is to achieve a very dense 3D packing of optically active units, which can then be used to build a quantum dots laser [297]. A lower threshold current, temperature stability, and a high gain in lasing and light output are some of the advantages of the QD laser, which have their origin in the discreteness of the DOS [298]. However, in order to achieve an overall high light intensity it is necessary to work with a large number of QDs, which are densely packed, of high crystalline quality, and exhibit a very narrow size distribution. The requirements for the QD synthesis are therefore quite stringent, and the formation of QD superlattices is one step in the direction of producing the requisite densely-packed QD ensembles.

The control of in-plane (horizontal) positioning is still a challenge, although it can be achieved by a combination of lithography-like techniques with Stranski–Krastanov growth: the surface is modified prior to growth through the introduction of preferential nucleation sites which can be achieved by writing damaged sites with a focused ion beam [297]. The vertical alignment of QDs, however, is achieved by the spatial modulation of the strain field imprinted on a semiconductor layer, which is grown as a cap layer on the QD layer [283, 299–301]. This cap layer is then the substrate for the next QD growth sequence and defines the spacing between the QD layers. The vertical ordering of QDs in Stranski–Krastanov growth is due to strain, which is a single, dominant ordering parameter compared with the multiple types of interactions, which drive the formation of colloidal QD superlattices.

The coupling between the adjacent QD layers can be adjusted by the thickness of the semiconductor interlayer [283, 302]. The placement of the QDs in the second and all subsequent layers is determined by subtle local differences in strain, which are induced by embedding the first layer QDs in the semiconductor matrix. A cross-sectional image of such a QD superlattice is shown in Figure 5.12. The variation in strain within the semiconductor interlayer functions as the ordering parameter for the vertical alignment between the individual QD layers but at the same time the size distribution of QDs changes from layer to layer: small QDs in the first layer will introduce insufficient levels of strain to initiate the nucleation of a QD in the subsequent layer, and therefore only larger QDs can propagate vertically. The QD distribution therefore narrows, and Figure 5.13 illustrates the evolution of the size distribution in a Si–Ge superlattice [283]; similar ordering is observed for III–V quantum dot superlattices. These QD superlattices provide superb control over the size and spatial distribution of QDs and a very dense 3D packing.

Xie *et al.* [301] systematically studied the degree of second layer ordering as a function of the semiconductor interlayer thickness for InAs QDs. A thicker cap will lead to a smaller variation in the surface strain field and therefore the vertical correlation of QD positioning is diminished, and disappears at a

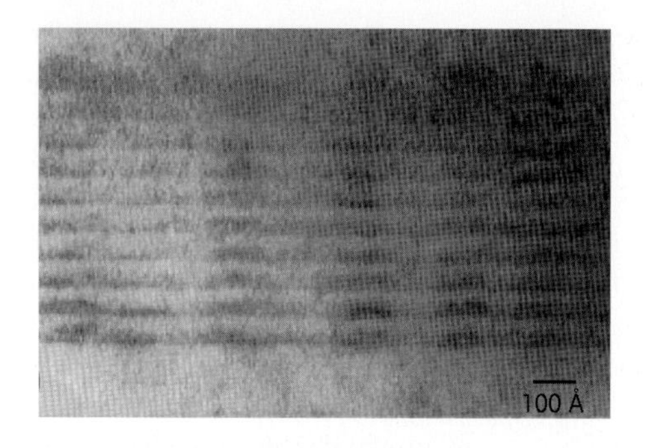

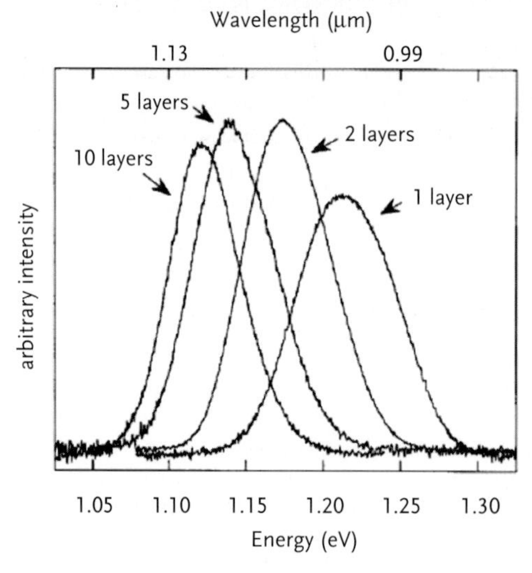

Figure 5.12 TEM image of a QD superlattice with InAs QDs embedded in a GaAs matrix [300]. The interlayer thickness is given as 5.6 nm, and the alignment of the QDs can be seen very clearly. The graph on the bottom illustrates the change in the photoluminescence spectrum (recorded at 8 K) as a function of the number of layers integrated in the superlattice. The narrowing of the linewidth is attributed to the narrowing of the QD size distribution, and the spectral shift is presumably caused by QD coupling and miniband formation, and a change in the QD average size. Reprinted figure with permission from Phys. Rev. Lett. [300]. Copyright 1996 by the American Physical Society.

material specific critical interlayer thickness. This loss of registry between the QD layers was attributed at least partially to a modification of the ad-atom diffusion in the early stages of QD growth [302]. The ad-atoms diffusion is assumed to consist of two major components: a thermally-driven diffusion, which is independent of the semiconductor interlayer thickness, and diffusion due to a mechanochemical potential gradient, which is controlled by the local strain field and therefore a strong function of interlayer thickness. If the strain field is then reduced with increasing interlayer thickness, the strain-driven diffusion will be greatly reduced

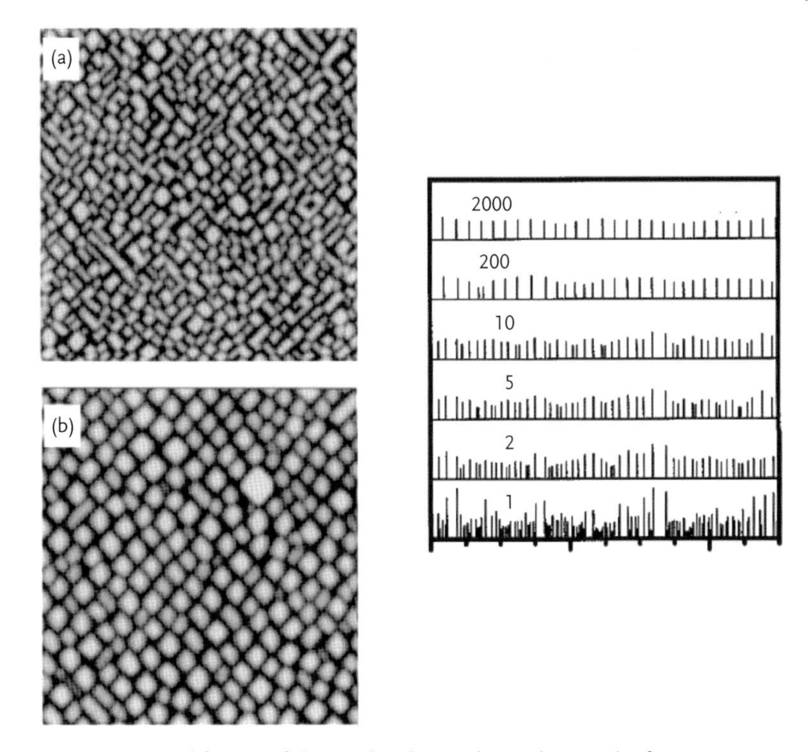

Figure 5.13 Modification of the QD distribution during the growth of a QD superlattice [283]: (a) is the first QD layer of $Si_{0.25}Ge_{0.75}$ QDs on Si(100), and (b) the QD distribution after the growth of a total of 20 layers. The AFM image size is (a) 0.8^2 μm^2, and (b) 1.2^2 μm^2. The graph on the right hand side illustrates a simulation the QD size evolution as a function of the number of layers; each line represents a QD and the height is proportional to the QD volume. The focusing of the distribution is already apparent after only a few layers, and a nearly perfectly uniform distribution is obtained after 200 layers. 2000 layers is a thickness, which is experimentally out of reach. Reprinted figure with permission from Phys. Rev. Lett. [283]. Copyright 1996 by the American Physical Society.

and the growth is entirely controlled by the thermally driven diffusion. While this model is likely an oversimplification of a complex nucleation and growth process, it illustrates the impact of local strain fields on growth.

However, the thickness of the semiconductor interlayer also plays an important role in the adjustment of electronic coupling between the individual QD layers. If the interlayer thickness is sufficiently small and in the range of at most a few nanometers, the individual QDs can interact, the wavefunctions of electrons and holes overlap, and charges can tunnel between the dots [300]. The QDs are no longer electronically isolated entities and the wavefunction overlap leads to the formation of minibands, a deviation from the perfect 0D DOS.

5.5
Closing Remarks

Quantum dots are an incredibly rich area of research: the variability in composition, size, and shapes appears nearly limitless. The next step is to improve our ability to assemble quantum dots into superlattices, both from colloidal and strain-engineered dots, and thus move from artificial atoms to artificial solids. The synthesis of nano-metamaterials where a variety of nanomaterial building blocks are combined to fabricate materials with new properties, and combination of properties, is going to be an exciting area of research.

6
Pure Carbon Materials

6.1
Carbonaceous Materials and Bonding

The unique structural versatility of carbon is based on its ability to adopt three different states of hybridization, namely sp^1, sp^2, and sp^3, which determine the local geometric arrangement around the carbon atom. The free carbon atom has an atomic orbital ground state configuration of $1s^22s^22p^2$ but since the stability of a chemical bond can be directly related to the overlap of electronic orbitals participating in the bond, it is favorable to maximize the overlap through hybridization of the 2s and 2p orbitals. The hybridization constitutes an excited state in the free atom and only becomes favorable upon bond formation. The type of hybridization determines the connectivity of the resultant structure, where the directionality in bonding is provided by the formation of covalent bonds. While the sp^1 hybridization allows the formation of linear chains and triple bonds between the atoms (two π- and one σ-bond), sp^2 hybridization leads to the formation of two-dimensional (2D) graphene sheets with a local trigonal-planar bonding environment and double bonds between the carbon atoms (one π- and one σ-bond). A three-dimensional (3D) network can be formed from sp^3-hybridized, tetrahedrally coordinated carbon atoms and is the basis of diamond, the hardest material known to date. In addition to the covalently bonded networks, van der Waals forces between single graphene sheets and the deviation from the perfectly planar graphene geometry yields π-bonded materials, such as fullerenes, carbon nanotubes and graphene, which have revolutionized nanoscience.

Graphite and diamond occur naturally and under ambient conditions graphite is the thermodynamically stable phase [303]. Although cubic diamond is the most well-known allotrope, hexagonal diamond (Lonsdaleite) has been reported as a high pressure phase, and has been found in meteorites [304, 305]. The same holds true for graphite, which is usually found in a hexagonal lattice but a different stacking of the graphene sheets can lead to a rhombohedral symmetry [306, 307]. The structural diversity of

Inorganic Nanostructures: Properties and Characterization, First Edition. Reinke, P.

Published 2012 by WILEY-VCH Verlag GmbH & Co. KGaA

pure carbon materials is not limited to the crystalline phases and also includes amorphous structures, with varying contributions of sp^3- and sp^2-hybridized carbon atoms (the sp^1 hybrid plays only a minor role in this context). The formation of amorphous structures can only be achieved if deposition methods far from thermodynamic equilibrium are used, such as plasma- or ion-assisted methods.

The variety in pure carbon structures can be visualized in the structural phase diagram shown in Figure 6.1. Diamond and graphite are the pure, crystalline sp^3 and sp^2 allotropes, and are on the left-hand side of the phase diagram, the amorphous phases for both hybridizations are on the right-hand side, and the center of the diagram is occupied by materials which contain crystalline and amorphous components and are phase mixtures of sp^2 and sp^3 carbon atoms. The latter are usually thin-film materials, and their mechanical, electronic, and optical properties vary widely. Although these thin film materials are usually called amorphous, their microstructure, short and medium range order, are still subject to discussion [308]. This is also related to the difficulty of obtaining experimental data, and the term amorphous has to be seen in connection to the respective experimental method used for structural analysis. If the processing parameters are varied, nearly the entire phase space can be accessed: sp^2-amorphous thin films are produced by the condensation of carbon atoms with low kinetic energies, or through the destruction of graphite or

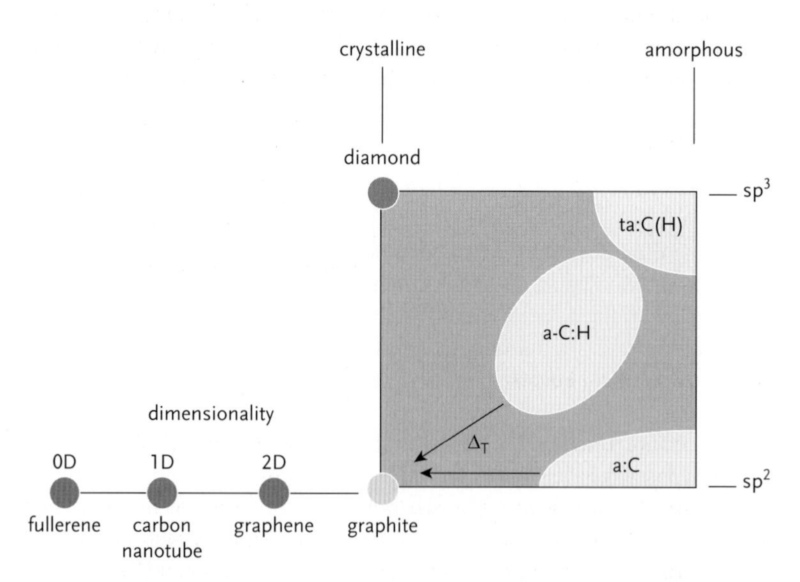

Figure 6.1 Structural phase diagram of carbon. The square includes crystalline and amorphous 3D structures with different contributions from sp^2 and sp^3 hybridized carbon. The transition from any amorphous phases to graphite can be achieved by moderate heating to temperatures around 800 °C. The third axis (on the left-hand side) denotes the dimensionality of the carbon structure.

diamond by ion irradiation; sp^3-amorphous thin films are produced when the ion energy is sufficient to displace atoms in the solid during thin film growth from hydrocarbon precursors, but insufficient to allow for a full relaxation of the structure [309]. The film is compacted and thus the sp^3-type coordination is favored. A-C : H films, which are often also called diamond-like carbon due to their extraordinary hardness, include hydrogen in a the amorphous network of sp^2- and sp^3-hybridized carbon atoms. The ratio of sp^2 to sp^3 carbon determines the hardness of these films, and it can be controlled by the processing parameters (ion energy, hydrogen concentration, temperature [310]). These films have been proven very useful as protective and optical coatings. Transitions between different structures can be initiated by post-deposition treatments such as ion irradiation or annealing: annealing will always push the system toward the thermodynamically stable graphite phase, whereas ion irradiation favors highly amorphous structures. Ordering within the plane of the graphite honeycomb lattice occurs first, and is followed at much higher temperatures with ordering along the c-axis perpendicular to the graphite sheets. Diamond [311] itself will begin to graphitize from the surface inwards at around 1250 °C.

6.2 Low-Dimensional Carbon Nanostructures

The low-dimensional carbon nanostructures, which include fullerenes, carbon nanotubes and graphene, are all dominated by sp^2 hybridized carbon atoms and positioned on the left-hand side of this phase diagram. We will begin this section with a description of the geometry of the carbon nanostructures, which is surprisingly versatile despite using sp^2-hybridized carbon atoms as the sole building block. This versatility can be traced back to the considerable flexibility of the trigonal planar geometry, which can accommodate deviations from planarity, which are sufficient for the formation of spherical or round objects such as fullerenes and nanotubes.

6.2.1 Zero-Dimensional – Fullerenes

Fullerenes are often considered to be zero-dimensional (0D) materials, but they possess a delocalized π-bonded, electronic system and might be more correctly classified as organic molecules. C_{60} was the first fullerene molecule to be isolated in sufficient amounts to perform structural analysis and confirm its unique spherical shape. H. Kroto, R. Smalley and R. Curl [312] were awarded the Nobel prize in chemistry in 1996 for the discovery of C_{60}, which was the first molecule of an entirely new class of materials called fullerenes. The name fullerene is derived from Buckminster Fuller, an architect and engineer who perfected the design of geodesic domes, such as the one constructed for the US pavilion for Expo 67 in Montréal. The fullerenes are a class of carbon molecules, which are closed cage polyhedra made of n sp^2 hybridized, threefold coordinated carbon atoms. The polyhedra consist of 12 pentagonal faces, and (n/2-10) hexagonal faces with $n \geq 20$, and their structure adheres to Euler's rule of polyhedra, which rules the formation of closed cage structures made of polyhedra. C_{60} is one of the most stable fullerenes, and a wide range of larger and some smaller molecules have been synthesized. The pentagons in most stable fullerenes are surrounded by five hexagons, and therefore isolated from the other pentagons (isolated pentagon rule, IPR). Figure 6.2 shows the C_{60} molecule (**1**) and a larger IPR fullerene where the pentagons are clearly visible in the cage structure (**2**). The electrons are delocalized and form a π-bonded, electronic system, which is nearly aromatic. The degree of overlap between adjacent p-orbitals is slightly less than in benzene, which reduces the molecules aromaticity. The presence of hexagons and pentagons leads to nonequivalent carbon atom positions, which are distinguished by their chemical reactivity [313–315].

In addition to the pure material, a variety of doped fullerene solids have been produced: the endohedral doping is the encapsulation of one or several atoms within the cage, while exohedrally-doped fullerenes

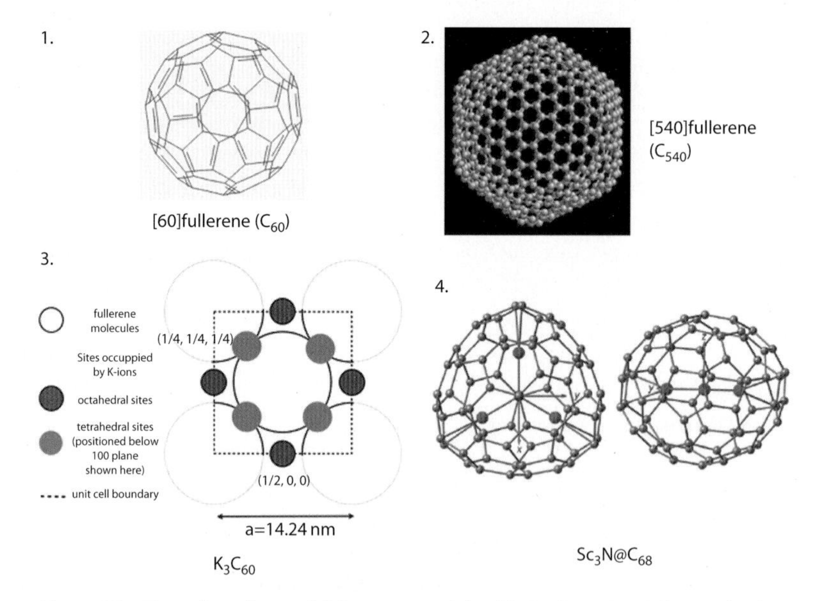

Figure 6.2 Examples of several fullerene materials: (1) C_{60} [http://pubchem.ncbi.nlm.nih.gov/], (2) C_{540} [http://en.wikipedia.org/wiki/File:Fullerene_c540.png], (3) K_3 C_{60} – the determination of the structure of this exohedrally doped fullerene material is described in [466], (4) $Sc_3N@C_{68}$ is a non-IPR fullerene and the slightly larger spheres are the Sc atoms with an N atom at the center [465]. Figure (4) reproduced with permission Chem. Eur. J. [465] John Wiley and Sons, Copyright 2006.

are akin to graphite intercalation compounds, where the dopant is positioned in between the fullerene molecules. One example for an exohedrally doped fullerene solid K_3C_{60} [466] and an example for an endohedrally doped fullerene $Sc_3N@C_{68}$ [465] are included in the figure. More examples on endohedral fullerenes are given in [316, 317, 318]. These endohedral fullerenes are synthesized by introducing the respective metal in the gas phase/carbon electrodes during fullerene synthesis, and then isolated by chromatography. In K_3C_{60} the metal (M) is positioned on the octahedral and tetrahedral sites of the fullerene lattice (f.c.c.).

The chemistry of fullerenes is very rich and has grown into a subspecialty of organic chemistry, where the fullerenes are used as templates for the attachment of a wide range of organic molecules. The functionalization of fullerenes is essential to increase their solubility so they can be incorporated in organic photovoltaic cells [319], where they act as acceptor material, and in nanomedicine, where they have become an important vehicle for the transport of biologically active agents [320]. However, while fullerenes have been studied intensely and are now incorporated in several commercial products, their toxicity is, like for so many other engineered nanomaterials, only poorly understood. The question of fullerene toxicity moved to the front page several years ago after publication of a study about the very adverse effects of fullerenes on the health of fish [321]. This study led to an intense discussion of the validity of the severity of this claim, and initiated several studies into the toxicity of fullerenes [322, 323] and more generally, nanomaterials. The impact of the increasing volume of manufactured nanomaterials on the environment, human and animal health is still poorly understood, and is now recognized as an important and integral part in the manufacture and use of nanomaterials [324].

6.2.2 One- and Two-Dimensional – Carbon Nanotubes and Graphene

The one- and two-dimensional pure carbon nanostructures are carbon nanotubes and graphene, which is a single sheet within graphite's layered structure. The geometric structure of nanotubes is derived directly from the graphene honeycomb lattice, which is shown in Figure 6.3. According to the IUPAC definition (International Union of Pure and Applied Chemistry[1]) graphene is "A single carbon layer of the graphite structure, describing its nature by analogy to a polycyclic aromatic hydrocarbon of quasi infinite size. Because graphite designates that modification of the chemical element carbon, in which planar sheets of carbon atoms, each atom bound to *three neighbors in a honeycomb-like structure,* are stacked in a

[1] http://goldbook.iupac.org/G02683.html

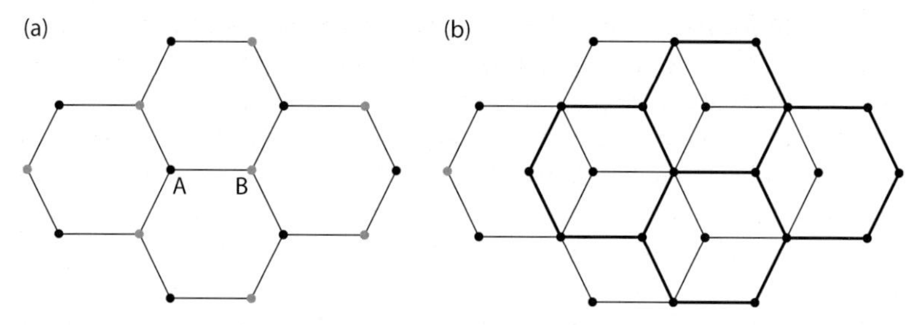

Figure 6.3 (a) The honeycomb lattice with its two nonequivalent lattice atoms A and B. (b) Superposition of two honeycomb lattices in a A-B stacking sequence as it is realized in graphite. Only one of the lattice atoms has a direct bonding partner in the next layer.

three-dimensional regular order, it is not correct to use for a single layer a term, which includes the term graphite, which would imply a three-dimensional structure. *The term graphene should be used only when the reactions, structural relations or other properties of individual layers are discussed.*" We will adhere to this definition and denote only the single layer as graphene. Two layers of graphene are a bi-layer, and several layers are called few-layer graphite (FL graphite). Graphene is the basic building block of graphite, which is made of stacks of graphene layers. Two different stacking sequences are known, and emerge from nonequivalent positions of the honeycomb lattices with respect to each other: the most common stacking is the so-called Bernal stacking with a ABABA . . . sequence, where the second layer is shifted in such a way that half the atoms from layer 1 are positioned on top of half the atoms in layer 2 (Figure 6.3) and the much rarer rhombohedral ABCABCA . . . stacking. The graphene layers are bonded to each other by van der Waals bonds, and possess a delocalized π-system within the layer. FL graphite already behaves like graphite for four or five layers of graphene if these are aligned in the proper Bernal stacking sequence. However, a small rotation of the graphene layers with respect to each other can lead to electronic decoupling, and the layers in FL graphite can behave like a stack of individual graphene sheets [325].

6.3 Electronic and Geometric Structure: Graphene and Carbon Nanotubes

In order to understand the origin of the carbon nanotube electronic structures we need to first take a closer look at the band structure of graphite, which is the building block for all nanotubes. The nanotubes can be considered as "geometrically" constrained graphene, and their electronic (and phononic) band structures are derived from the graphene building block. Graphite and graphene are probably currently the best-known and most studied layered material, but a wide range of solids present a similar geometric structure. Boron nitride for example has a hexagonal phase, which is isoelectronic to graphite, albeit the B–N bond is much more ionic and the BN layers are therefore insulators with a relatively large band gap [326–329]. MoS_2, which is a well-known lubricant, is also a layered compound in the family of layered transition metal dichalcogenides [330]. Most of these materials have been studied as bulk solids, and for some of them tubular structures equivalent to carbon nanotubes are known. While they have not played a large role in nanoscience and nanotechnology applications so far, they are very likely to attract more interest as we start building 2D and 1D device structures, which are assembled solely from nanoscale building blocks.

6.3.1 From Graphene to Graphite to Graphene

It is quite remarkable that graphene is a rather old material, at least in the context of theoretical calculations of its band structure. One of the most important publications on graphene (then still called 2D graphite, or single layer graphite) was written by Wallace [331] in 1947, and is a tight-binding calculation of the electronic band structure of single and multilayer graphite. However, it was inconceivable at the time that single sheets of graphene could be isolated, and the use of graphene was considered an easy path to simplify a rather complex 3D structure. The computational challenges arose mainly from the difficulties in describing the long-range van der Waals interactions. Painter and Ellis [332], and later Zunger [333, 334] presented *ab-initio* calculations of the electronic and optical properties of graphene and graphite in 1970, which used a variational approach, and an LCAO basis set (linear combination of atomic orbitals). The band structure of graphene [335] is depicted in Figure 6.4, and it can easily be seen that the occupied π and unoccupied π^* bands, which are the valence and conduction band, meet at the K-point in the graphene Brillouin zone.

When graphene is discussed we usually limit our attention to the section of the band structure, around the K-point, which drives graphene's extraordinary properties. The symmetry of the band, the linear slope, and the zero overlap of the valence and conduction band are the most prominent features, and we will return to them in the section on graphene. The σ-band, which originated from the covalent carbon–carbon bond between the hybridized orbitals lies deeper in the band structure, but crosses the π-band at the center of the Brillouin zone. When additional layers of graphene are added, the change in the overall band structure appears deceptively small: the curvature of the bands changes slightly, and more significantly, the valence and conduction

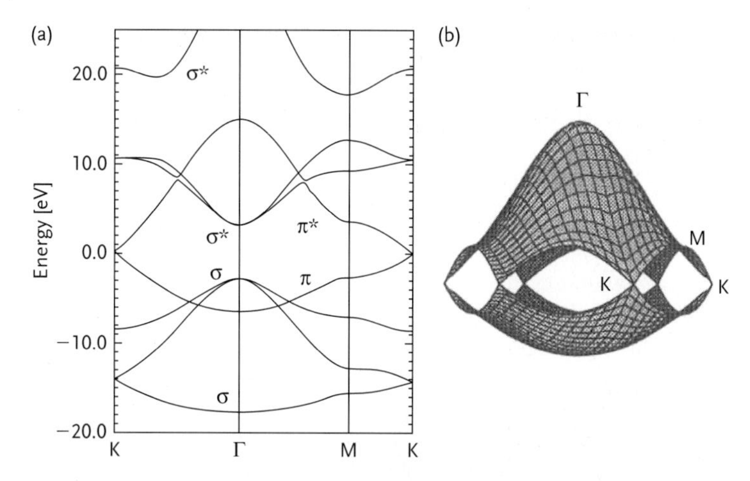

Figure 6.4 (a) Band structure of graphene [335] with the π and σ bands; (b) a small section of the band structure around the high symmetry points in the Brillouin zone K, and M. This depiction is often chosen in the discussion of graphene, and omits the deeper lying σ bands, and the crossover of the two bands close to the center of the Brillouin zone. Reproduced with permissions from [335].

bands now overlap at the K-point thus making graphite a semimetal [337–339]. The band structure of FL graphite moves through the transition regime, and therefore exhibits a complex mixture of graphene and graphite characteristics.

6.3.2
Geometric Structure of Carbon Nanotubes

The nanotube structure can be visualized by printing the hexagonal honeycomb graphene lattice on a sheet of paper and exploring the many possibilities of constructing a closed tube where the bonding geometry of the carbon atoms is conserved around the perimeter. Note that the circumference and rolling of the sheet can only be achieved in discrete units since the bonding geometry of the carbon atoms needs to be conserved [340–344]. The unit cell of the honeycomb lattice is indicated in Figure 6.5 with the basis vectors $\vec{a}_1$ and $\vec{a}_2$. When the honeycomb lattice is now "rolled" and the carbon nanotube is formed by connecting the origin O to point O^*, and B is connected to B^* while preserving the carbon atom bonding geometry. The vectors $\vec{T}$ and $\vec{C}_h$, which is also called the chiral vector, are along the tube axis and perpendicular to the tube axis, respectively as shown in Figure 6.6. These vectors fully describe the tube structure and define the nanotubes unit cell. The number of hexagons N in a nanotube unit cell is given by the area of the unit cell divided by the area of the graphene unit cell, the

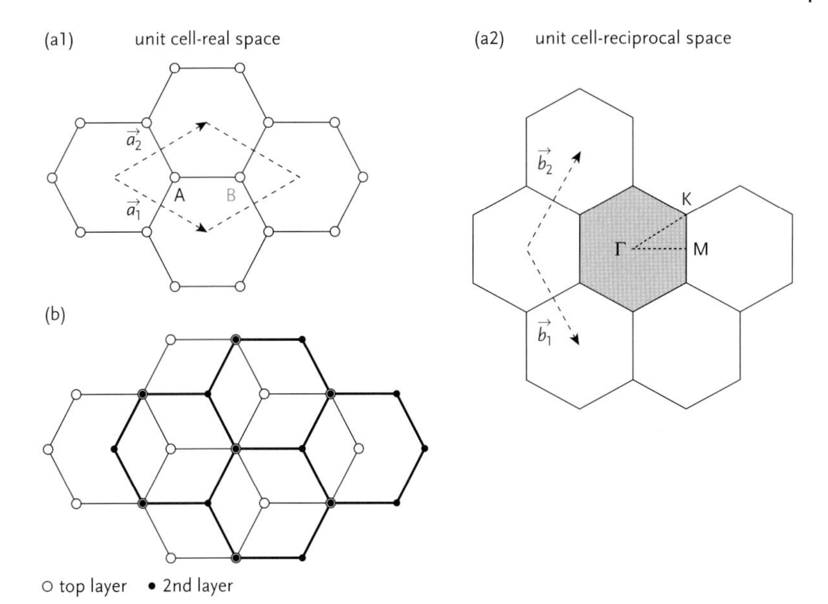

Figure 6.5 Unit cell of the honeycomb lattice in real space (a1, b) and in reciprocal space (a2). The Brillouin zone is marked in gray, and contains the origin Γ and the K and M points at the boundary of the Brillouin zone.

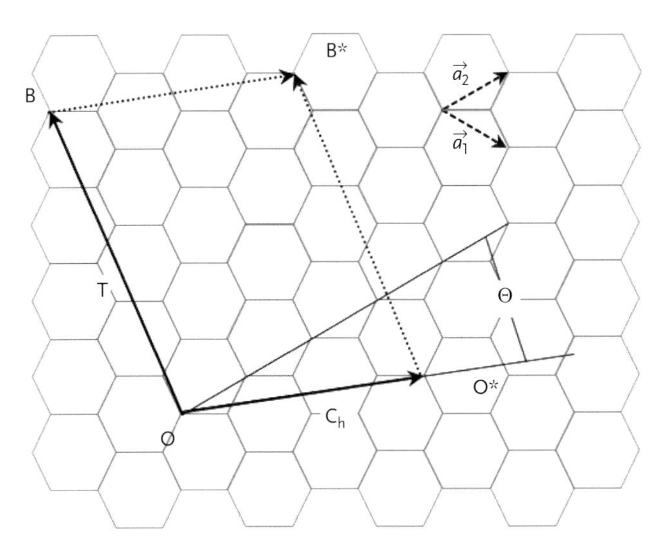

Figure 6.6 Construction of the unit cell of a carbon nanotube. The unit cell of graphite is indicated in the upper right-hand corner of the honeycomb lattice, and the nanotube unit cell is defined by the vectors C_h and T.

number of electrons per nanotube unit cell is then $2N$, and the number of π bands is then N.

The relation of the carbon nanotube unit vectors to the unit cell vectors of graphene is given by:

$$\vec{T} = t_1\vec{a}_1 + t_2\vec{a}_2 \quad (6.1)$$

$$\vec{C}_h = m\vec{a}_1 + n\vec{a}_2 \quad (6.2)$$

where t_1, t_2 and m, n are integers. It should be noted that the nanotube unit cell can be quite large and contains consequently a large number of electron leading to an equally large number of electronic and phononic bands. The chiral angle Θ is defined by the smaller angle enclosed by $\vec{C}_h$ and $\vec{a}_1$. The rolling of the nanotube leads for Θ of 0° or 30° to two different arrangements of the hexagons if seen along the tube axis: the armchair and zigzag configurations. An armchair configuration is achieved for $m = n$ and has a chiral angle of 30°, and a zigzag tube is obtained for $m = 0$ with a chiral angle of 0°. These chiral angles lie on the sixfold symmetry axis of the basic hexagon and the tubes will therefore be achiral. All other tubes exhibit a so-called "handedness" and are chiral tubes. The chiral angle is given by: $\arccos\,(n + m/2)/(n^2 + nm + m^2)$ [335]. A left- and a right-handed tube can then be constructed for each combination of m and n, and these tubes are mirror images of each other.

6.3.3 Electronic Structure of Carbon Nanotubes

One of the most fascinating properties in carbon nanotubes is the close relation between their geometric and electronic structure. The geometry fully defines whether it is a semiconducting or a metallic tube. This can be understood in a qualitative manner by looking at the electron wave function, which spans the circumference of the nanotubes. The nanotubes circumference as defined by the chiral vector $\vec{C}_h$ imposes an additional boundary condition on the electron wave, and the wave vector $\vec{k}$ is confined and obeys the relation $\vec{C}_h \cdot \vec{k} = j2\pi$, where j is an integer. This is akin to the formation of sub-bands, which was discussed in the context of semiconductor nanowires in Chapter 3. The energy is quantized in one direction, namely around the tube circumference, but not quantized in the other direction along the tube and the sub-bands can be represented by lines in the 2D k-plane of the dispersion relation $E(k)$ of graphene (the third axis is the energy). When a sub-band of the radially confined graphene electronic states crosses the Fermi surface of the graphene honeycomb lattice, the carbon nanotube will be metallic. This approach is the called zone folding, and it uses solely the $\pi\,\pi^*$ band of graphene. The zone folding can be derived in a more qualitative manner by examining the Brillouin zones (BZ) of graphene, and those of the carbon nanotube [340–342]. The Brillouin zone of the honeycomb structure is also a hexagon, and is shown in

Figure 6.5. For graphite (and graphene) the K-point in the BZ marks the Fermi surface where the valence and conduction bands touch. A metallic tube is created if the sub-band created through the confinement along the tube perimeter coincides with the Fermi surface of the graphene honeycomb lattice.

The lattice vectors $\vec{b}_1$ and $\vec{b}_2$ in the reciprocal lattice are derived from $\vec{a}_1$ and $\vec{a}_2$ using the relation given in solid state textbooks. The real space lattice vectors C_h and T of the nanotube are described in terms of the graphene lattice unit cell vectors.

The reciprocal lattice vectors $\vec{k}_1$ and $\vec{k}_2$ belong to the carbon nanotube unit cell, and are around the circumference, and along the axis, respectively. They relate to the reciprocal lattice vectors of the graphene lattice as following:

$$\vec{k}_1 = \frac{1}{N}(-t_2\vec{b}_1 + t_1\vec{b}_2), \text{and } \vec{k}_2 = \frac{1}{N}(m\vec{b}_1 - n\vec{b}_2), \tag{6.3}$$

where N is the number of hexagons in a nanotube unit cell. Figure 6.7 shows the relation between the Brillouin zone of a C_h (n, m) with $n = m$ nanotube and the graphene Brillouin zone, and for a C_h *(4,2)* tube.

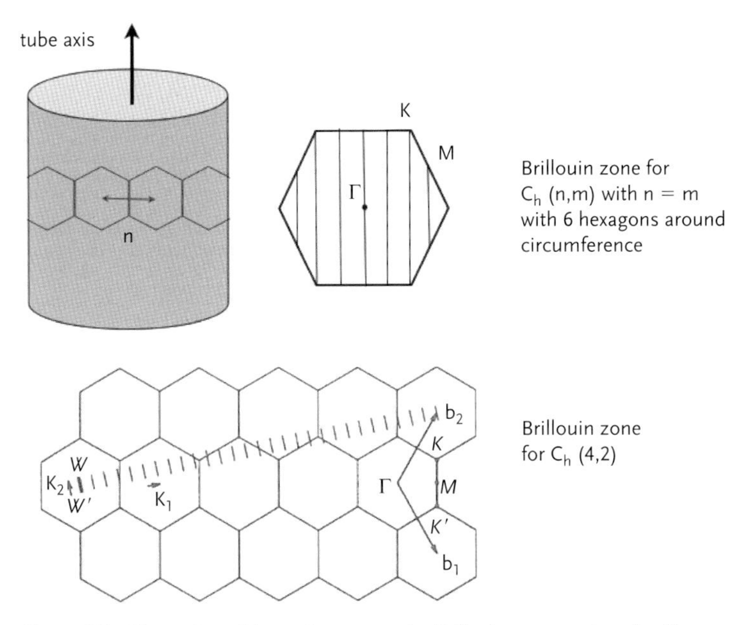

Figure 6.7 Illustration of the carbon nanotube Brillouin zones: a (n, m) with $n = m$ nanotube and six hexagons around the perimeter, and another example at the bottom for a (4,2) nanotube [335]. Note the tilt between the nanotube Brillouin zone (line segment) and the hexagonal reciprocal lattice of graphene. The $(n = m)$ tube is metallic since the nanotube Brillouin zone crosses the K-point in graphene, the (4,2) tube will be semiconducting since the nanotube Brillouin zone does not cross the K-point. The (4,2) tube was reproduced with permissions from [335].

The tilt angle between the axis of nanotube's BZ (the connecting line between the symmetry points W and W′) and the graphene's unit cell represents the chiral angle. If the carbon nanotube BZ is translated throughout reciprocal space by $\vec{k}_1$ it will either cross the K-point in graphene's reciprocal lattice, in which case the nanotube is metallic, or it will not cross the K-point and the tube is semiconducting. An assessment of the geometry of both BZs and their relative position yields that if the length of the connecting vector, which corresponds to the shortest distance between the nanotube BZ (W-W′) and the K-point in the honeycomb lattice BZ is a multiple of $\vec{k}_1$the tube will be metallic. If this condition is expressed in unit cell dimensions a tube will be metallic if $(2n+m)$ is a multiple of 3. The complete derivation of the geometric relation is given in the book on carbon nanotubes by Saito *et al.* [335]. The metallic and semiconducting tubes can therefore be identified from their unit cell vectors.

The relation between geometry and structure has been confirmed by numerous scanning tunneling microscopy (STM) and scanning tunneling spectroscopy (STS) measurements. The experimental challenge in this context is to immobilize the carbon nanotube and achieve atomic resolution on a curved surface, which then allows to determine (m, n). A local technique is required for this measurement, since it remains to this day difficult to synthesize large batches of well-defined types of carbon nanotube. The large variability in geometry and thus properties gives an unprecedented breadth for the design of carbon-based nanoscale devices, but the lack of a reliable and highly selective synthesis is still a considerable problem. Figure 6.8 shows STM images of two different nanotubes, one metallic and one semiconducting tube, and the corresponding STS data and calculations of the density of states (DOS) [336]. The band gap is identified by the zero current around the Fermi energy, and the sharp peaks in the conduction and valence bands are van-Hove singularities. They appear at each point in the DOS where dE/dk becomes zero, and consequently the DOS diverges. This is seen in the depiction of the DOS for the 1D system discussed in Chapter 1, and in the DOS (E) for the carbon nanotube included in Figure 6.8. The development of the band gap as a function of size for carbon nanotubes and semiconductor nanowires follows the same trend: in both systems the diameter of the 1D structure determines the band gap, and the band gap is reduced with wire/tube diameter due to the confinement. According to Saito *et al.* [335] and White and Mintmire [341] the band gap E_G is inversely proportional to the tube diameter d, while the tube chirality when we compare semiconducting tubes, plays a minor role.

The ability to tune the electronic structure of nanotubes leads us immediately to the next idea, namely the fabrication of on-tube device structures, where small sections are metallic or semiconducting and the

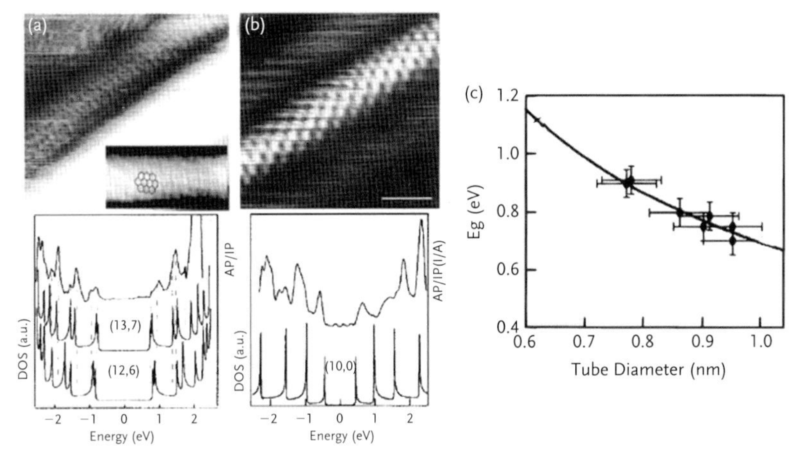

Figure 6.8 STM images of two different carbon nanotubes: (a) (13,7) tube which is assigned using the position of the van-Hove singularities in the DOS [336]. The comparison of STM image, band gap, and position of the singularities in the scanning tunneling spectroscopy with the tight binding calculation shown below the STM image allowed for the identification of the tube. This tube is metallic, which is apparent from the current in the dI/dV spectra and DOS around the Fermi energy at 0 eV; (b) STM image is a (10,0) tube and therefore semiconducting. The imaging conditions were a bias voltage of 0.55 V (13.7) tube, and 0.6 V (10,0) tube. The feedback current was 0.12 nA and the tubes were immobilized on a Au(111) surface; (c) the graph shows the band gap as a function of the tube diameter for semiconducting tubes. The line corresponds to a band gap proportional to $1/d$ and thus is in excellent agreement with the tight binding calculations by Saito *et al.* [335] and White and Mintmire [341]. Reprinted with permission from J. Phys. Chem. B [336]. Copyright 2000 American Chemical Society.

entire tube defines the respective device functionality [345–347]. It is indeed possible to see and measure metal–semiconductor junctions on single tubes, but it is difficult, and some would say even impossible, to control the fabrication of segmented tubes. The only way to switch between metallic and semiconducting is to change the pitch of the tube, however, this can only be achieved by the introduction of a structural defect in the form of pentagons or heptagons. The current–voltage characteristic of an on-tube metal–semiconductor junction has indeed been measured by placing a kinked tube between two contact pads. Figure 6.9 illustrates the tube configuration and shows the respective current–voltage characteristic. Kinks are usually introduced in the tube during growth, but can also be made artificially through electron or ion bombardment. This process, however, is often destructive to the tube structure, and most of the time this type of measurement is performed by probing a large number of tubes until one with the correct type of kink is identified.

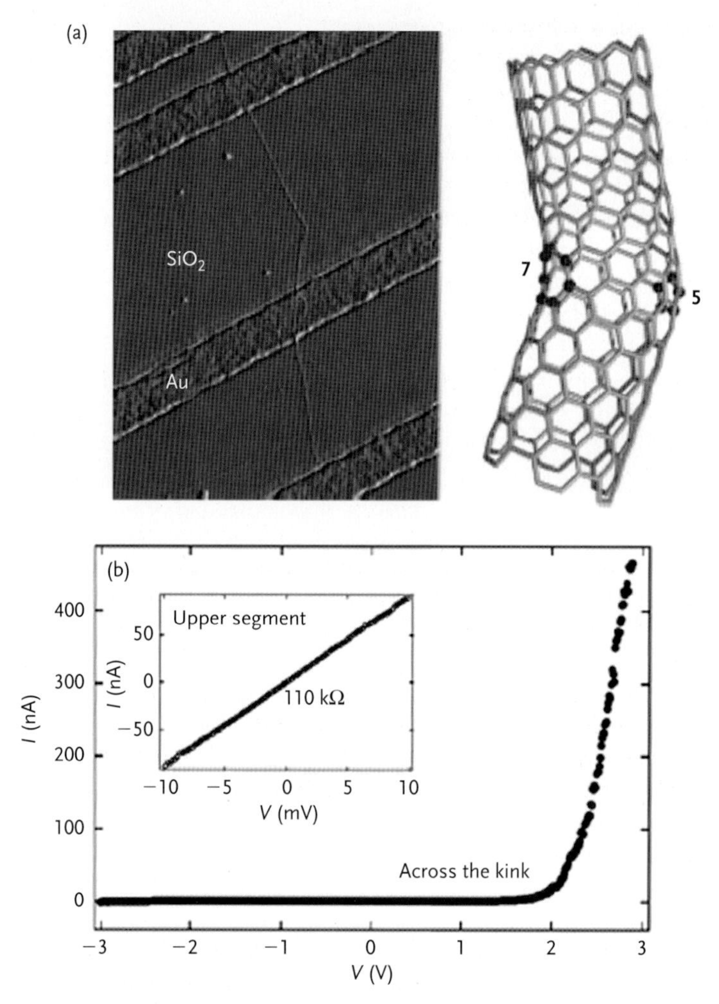

Figure 6.9 (a) The tapping mode AFM image shows a kinked nanotube, which is immobilized between two Au-contact pads [345]. The kink is formed by a pair of pentagon-heptagon rings within the tube, which modify the tube chirality and thus lead to the formation are a junction between a metallic and a semiconducting tube; (b) the *I–V* characteristics of the tube shown in the AFM image. The rectifying nature of the junction is confirmed by the flat region around the origin and the onset of conduction can be seen in the *I–V* curve. The I-V characteristics in the inset are measured in the straight (upper) segment) of the tube, where no structural defects are observed. Reprinted by permission from Macmillan Publishers Ltd: Nature [345], copyright 1999.

6.3.4 Synthesis of Carbon Nanotubes

In the last few sections we have talked about carbon nanotubes as an ideal, quite wonderful material, which can be described and understood by considering a rolled-up graphene sheet and its geometry. This kind of single wall carbon nanotube can, in analogy to Si, be considered a high-quality electronic material: it has a low concentration of defects, is structurally well-defined, and exhibits an intriguing combination of electronic properties. However, many applications for nanotubes do not require this superb crystallinity, but depend instead on the tube's highly anisotropic mechanical properties, its anisotropy in conductivity, the large surface for molecule absorption and sensing [346, 348–346]. Many of these applications are served well with tubes containing a larger number of defects, multi-wall tubes or even relatively thick carbon fibers.

Multiwall carbon nanotubes consist of several nanotubes, which are nested like a Russian doll, and can contain a large number of shells. The measurement of the crystallography of the multiwall tubes, their relative orientation and bonding between the tubes remains a very interesting and challenging problem. While electronic transport along the long axis of the tubes is well understood, transport across the multiple shells [348, 349, 347] remains a challenge and is presumed to depend largely on the inter-shell bonding, which can be of the van der Waals type and might allow the tubes to rotate with respect to each other, or may be covalent bonds between the shells, which can fix the tubes.

An ensemble of nanotubes is characterized, firstly, by the properties of the tubes: single or multiwall tube, diameter, length, and chirality, and secondly, by their geometric arrangement: vertical or horizontal, intertube distance (density), and attachment points. The envisaged application defines which ensemble characteristics are important and which are less critical to achieve the desired functionality. For example, to make a very black light-absorbing surface with reasonable thermal conductivity [354], the packing density of vertical tubes is the most important ensemble characteristics, the number of walls per tube, and the chirality of the individual tubes, however, is much less critical and probably does in this context not play a role. High density ensembles are characterized either as "mats" where the nanotubes are mostly vertical and the microscope images are reminiscent of interlinked spaghetti, or as "forests" which are long, free standing nanotubes with a high area density [355–359]. An example for a nanotube forest is shown in Figure 6.10.

Nanotube synthesis has been achieved by a variety of methods, including arc deposition, laser ablation, plasma-enhanced chemical vapor deposition, and chemical vapor deposition (CVD). They all have their merits, but the CVD methods currently dominate the market (and laboratory) since they afford superb control over processing parameters, large area deposition, control over tube diameter, and can be used with a wide range of samples [360–362]. Pure CVD techniques, which rely on catalyst-assisted decomposition of carbonaceous precursor, are likely to lead to a nondirectional growth, and the formation of mats. The plasma-assisted methods on the other hand can lead to the formation of extremely well-aligned nanotube

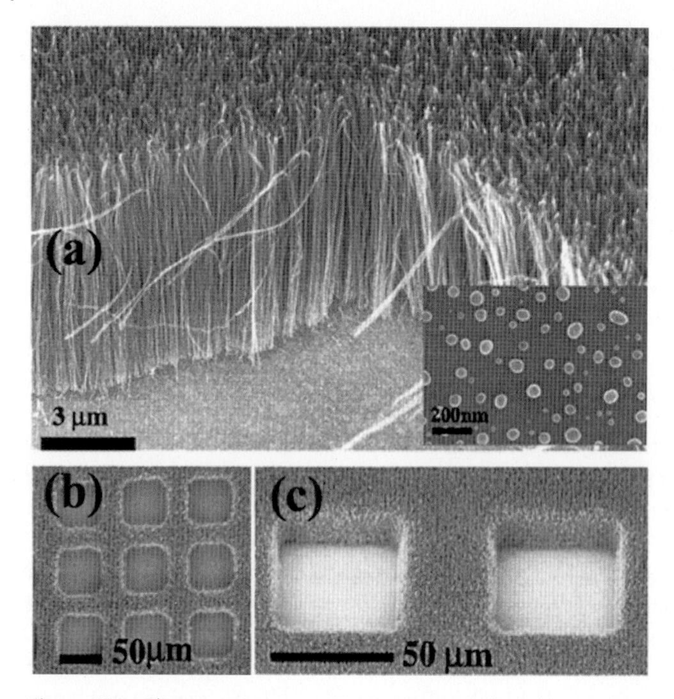

Figure 6.10 Electron microscopy (SEM) images of aligned nanotube "mats" [355] grown on Co-catalyst particles, which are shown in the inset in (a). Images (b) and (c) show samples where small square sections were not covered with catalyst particles, and are therefore bare. Reprinted with permission from Appl. Phys. Lett. [355]. Copyright 2000, American Institute of Physics.

forests, since the electric field at the plasma–substrate interface imparts a directional preference on the growth.

The mechanism for carbon nanotube growth is in many ways similar to the vapor–liquid–solid (VLS) process for semiconductor nanowire growth: metal catalyst particles with an appreciable solubility for carbon such as Fe, Ni, Co, or related alloys, are deposited on the substrate surface. The substrate is then heated and the catalyst exposed to a flux of carbonaceous precursors such as methane, higher hydrocarbons, or carbon monoxide (CO), which decomposes at the catalyst surface. The catalyst particles are then supersaturated in carbon, and precipitation of carbon triggers the growth of nanotubes. In contrast to semiconductor nanowires, the crystallographic relation to the substrate has, with few exceptions, no influence on the direction of nanotube growth. The lack of an epitaxial relation also means that the honeycomb lattice structure has to nucleate at the surface of the catalyst through precipitation of carbon. The growth of the nanotube can occur at the substrate catalyst interface, and the catalyst particle is then transported at the tip of the nanotube ("tip growth") like in VLS, or the tube grows out of the catalyst particle, which remains at the base of the tube and the substrate interface ("base growth"). The latter kind of tubes is closed at the end with a highly curved end-cap reminiscent of fullerenes, which can be removed by chemical etching. The type and presumably crystallinity of the catalyst and the substrate type as well as temperature of the process determine whether isolated, single-wall tubes, single-wall tube bundles, or multiwall tubes are the main product. The size of the catalyst particles, respectively the overall amount of catalyst, is directly proportional to the diameter of the nanotubes [356]. It has been suggested that chemical modification of the catalyst particle, such as the partial transformation to a carbide, local strain fields at the tube–catalyst interface, and the strength of the substrate–catalyst interaction determine the growth mode for nanotubes. The highest quality single wall tubes are nowadays fabricated with the high pressure carbon monoxide (HiPCO) conversion process, where Fe-carbonyl is used as catalyst precursor, and CO as carbon precursor [360, 363]. Many other processes used for nanotube synthesis still require purification to remove amorphous carbon and other side-products.

The density of nanotubes and their spatial distribution is in analogy to semiconductor nanowires, determined by the size, density, and position of the catalyst particles. The growth of carbon nanotube on quartz surfaces, which was used by Kocabas *et al.* [357] to fabricate building blocks for electronic devices, combines spatially selective catalyst deposition, and

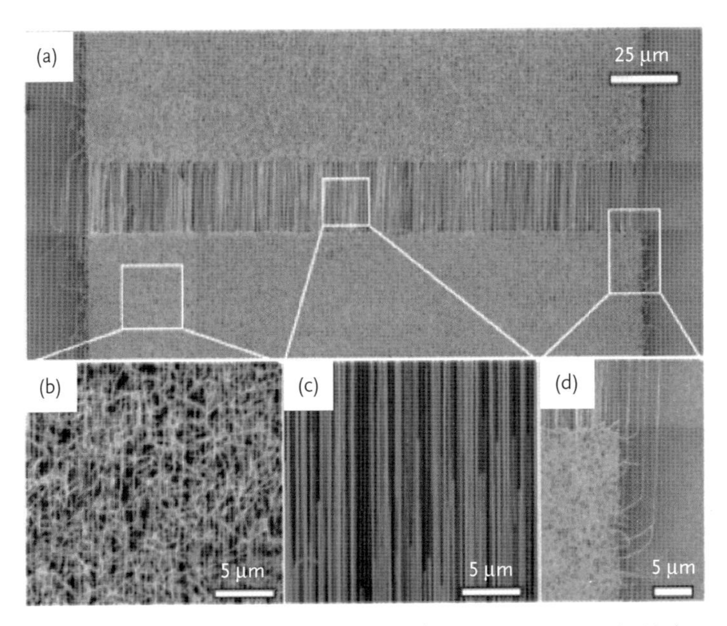

Figure 6.11 Nanotube arrays on a quartz surface [357] (SEM images): (a) the catalyst was added in patches above and below the region of well-aligned tubes in the center of the image. (b) shows the nanotubes grown in the region of the catalyst patches, (c) is the region without catalyst particles where the carbon nanotubes grow along preferred directions on the quartz surface, and (d) is the boundary of the nanotube array at the interface between aligned and non-aligend regions. Reprinted with permission from J. Am. Chem. Soc. [357]. Copyright 2006 American Chemical Society.

templated growth through the surface structure of a quartz substrate. An example for the growth on quartz is included in Figure 6.11. The goal of this study was to produce carbon nanotubes, which form a large array of parallel tubes instead of dense random arrays. The catalyst ferritin was deposited in two spots, defined by lithography, and the carbon nanotubes then grew along a well-defined crystallographic direction [2 −1 −10] from one catalyst patch across the surface to the next catalyst patch. In areas with a dense population of catalyst particles, more random growth was observed.

The use of carbon nanotubes for electronic application in circuits where metallic and semiconducting tubes with varying band gaps are coupled in a deterministic manner, however, requires the targeted fabrication of large batches of nanotubes selected by size, electronic structure, and chirality. None of the methods described above has yet been able to deliver monodisperse carbon nanotube populations. In recent years the manipulation of catalyst particle composition, size, and crystallinity has emerged as the most promising route to a highly selective growth process [364, 365]. For example, the use of a FeRu alloy leads to particularly small, stable catalyst particles, which allow the growth of nanotubes with diameters of only 1.2 nm where only few (m, n) combinations can be realized. The selectivity of the growth process appears to be closely related to very specific properties of the catalyst particles, and further investigations might indeed lead us to the desired selectivity during growth. However, the desired selectivity for electronic applications can currently only be reached by the combination of a selective growth process in combination with a post-growth separation process. A review of these techniques, which rely on the chemical reactivity of different types of nanotubes, is given in Ref. [365].

6.4
Graphene – the Electron as a Massless Dirac Fermion

The band structures of graphene and graphite were discussed at the beginning of this chapter, and we now turn our focus to graphene and its potential for future carbon-based electronics. Most of the unusual properties of graphene are derived from the shape of the bands around the Fermi energy at the K-point of the Brillouin zone. The slope of the valence and the conduction band are linear in $E(\vec{k})$ in a small energy window of approximately 1 eV around the K-point, which was predicted by Wallace [331] in his tight-binding description of the graphene band structure [366–373]. The valence and conduction bands are symmetric (ambipolar) and do not overlap but meet at a point, and no gap exists between them. The Fermi energy or chemical potential is positioned at the meeting point of the two bands. The electrons in the vicinity of the K-point can be described as massless, Dirac fermions: a quasiparticle, which exists in graphene because of the extraordinary geometry of the band structure. The electron bound in the honeycomb lattice now behaves as if it is a massless Dirac fermion; much like a bound electron in a lattice will not behave like a free electron with rest mass m_0 but like a quasi-particle with equal charge and an effective mass of m^*. It was recognized several decades ago that electrons within a honeycomb lattice, which is made of two sublattices A and B, can be described successfully within the Dirac framework, albeit for a long time this work remained a purely theoretical exercise [374]. The Fermi velocity of about 10^6 m s^{-1} for electrons in graphene takes on the role of the limiting speed of light in the Dirac description. The energy E is then given by $E = \pm\hbar v_F|\vec{k}|$, where $+$ and $-$ denote the conduction and valence band, respectively. For the derivation of the energy level spacing in the quantum confinement regime it is now necessary to use this relation for E instead of the one for the electron with an effective mass m^*.

The description of the electron as a massless Dirac fermion has considerable consequences for the behavior of graphene, which are not limited to the superb charge carrier mobility, but include a wide range of solid state physics phenomena. The synthesis of graphene allows us for the first time to test phenomena relatively easily, which have been predicted for particles described with the relativistic Dirac equation. The unique nature of graphene's electronic system has been confirmed by measurement of Shubnikov–de-Haas oscillations, which measure the cyclotron mass of charge carriers, the measurement of magnetic field dependence of the quantum Hall effect, the splitting of energy levels through the application of a perpendicular magnetic field (Landau levels), and transport measurements [370, 375–379]. In contrast to the single layer graphene, bilayer graphene presents an odd mixture of properties, and is described as a massive Dirac fermion with components from Schroedinger and Dirac descriptions. The $E(\vec{k})$ relation at the Fermi energy is already parabolic, but the bilayer graphene does not yet behave like the bulk material graphite [373, 380, 381]. As we will show in this text, the use of a bilayer graphene can be advantageous for the introduction of a small band gap, which might be easier to implement than for the single layer graphene.

It is tempting to compare the 2D graphene with a 2D electron gas (2DEG), a system, which is well-known from semiconductor nanostructures such as quantum wells. The critical difference between a 2DEG and the graphene honeycomb lattice are the boundary conditions imposed on the electron wave function. A 2DEG is described by building a standing wave, which is confined solely by the edges of the well where the potential energy barrier is very large. The electrons are treated as free or weakly bound electrons. The graphene lattice on the other hand is described by combining the electron wave functions with a Bloch function, which contains the information of the symmetry of the lattice. The tight-binding calculation for example uses a linear combination of atomic orbitals (LCAO) combined with the Bloch function to determine the band structure of graphene. The boundary conditions for the wave function in graphene are therefore given by the lattice geometry, which leads in a straightforward manner to the graphene's characteristic $E(\vec{k})$ relation for the π—band.

The band structure and DOS of graphene has been measured directly with angle resolved ultraviolet photoelectron spectroscopy (ARUPS) following the relation given in eq. 2.4, which yields the band dispersion for the entire Brillouin zone [382–384]. An example of the dispersion curve $E(\vec{k})$ as it is extracted from ARUPS measurements is shown in Figure 6.12 [377]. The linear slope of the bands, and the Dirac point where the valence and conduction band touch, can be identified with ease. However, the Fermi level, which separates the occupied and unoccupied states does not in this case coincide with the Dirac point, but is slightly shifted in energy, and the charge carrier additions to valence or conduction bands are akin to doping (Figure 6.13). This shift is due to a minute charge imbalance in the

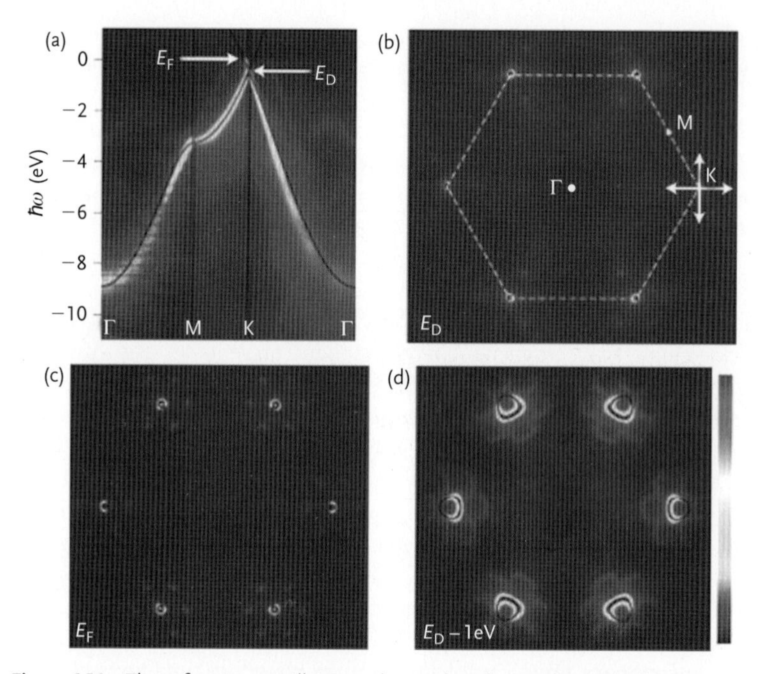

Figure 6.12 These four images illustrate the results of an angle resolved photoelectron spectroscopy study of the band structure of graphene layers [377] on SiC. (a) shows the band structure $E(k)$ of graphene measured along the principal directions in the Brillouin zone of graphene(white line), the dark lines are a band structure calculation described in the corresponding reference. The Dirac E_D and Fermi E_F energies are marked in the graph, and the shift of E_F with respect to E_D indicates a slight doping, most likely from the substrate; (b) is a constant energy map at E_D, and the arrows indicate the directions of data acquisition in the angle resolved measurement; (c) constant energy map at E_F; (d) constant energy map deeper in the band. For the constant energy map we take a "cut" through the bands mapped out in (a) at different energies with respect to the Dirac energy. The additional faint peaks around the K-point in the constant energy graph are due to low energy electron scattering at the reconstructed substrate/interface. Reprinted by permission from Macmillan Publishers Ltd: Nature Physics [377], copyright 2006.

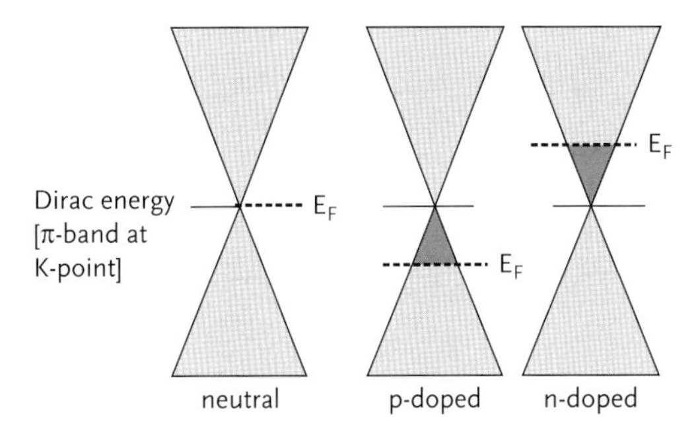

Figure 6.13 Schematic illustration of the Dirac cone in graphene, and the modulation of the position of the Fermi energy as a function of doping.

graphene layer, which can often be attributed to interactions with the substrate. In this particular case the substrate was SiC, and the graphene layer was synthesized by the sublimation of Si; this method is described in more detail in the next section. The influence of the substrate is one of the most critical issues in graphene synthesis and device fabrication, since it can significantly depress the highly coveted extraordinary mobility of charge carriers [376, 378].

6.4.1
Electronic Properties, Doping, and Band Gap

The electron transport in graphene is essential to its application in novel electronic devices, and for an ideal graphene layer the charge carrier transport should possess a sharp conductivity minimum for charge carriers located in the vicinity of the Dirac point [369, 376]. The general textbook expression for the conductivity σ is given by:

$$\sigma = n_e \boldsymbol{\mu}_e + n_h \boldsymbol{\mu}_h \tag{6.4}$$

and n_e, n_h are the charge carrier densities for electrons (e) and holes (h), and μ_e, μ_h are the respective charge carrier mobilities. In pristine graphene the charge carrier densities and mobilities for electrons and holes are identical due to the symmetry of valence and conduction bands around the Dirac point. The energy distance between Dirac and Fermi energy determines the charge carrier density, and exactly at the Dirac point the charge carrier density approaches zero and the conductivity reaches a minimum, which is shown in Figure 6.14. The conductivity is a function of a gate voltage, which shifts the Fermi energy relative to the Dirac point through the application of an electric field, therefore shows a V-shape with the same slope in conductivity on either side of the origin.

However, the conductivities which are observed for graphene on SiC or SiO_2 substrates are usually much lower than expected. The graphene sheets

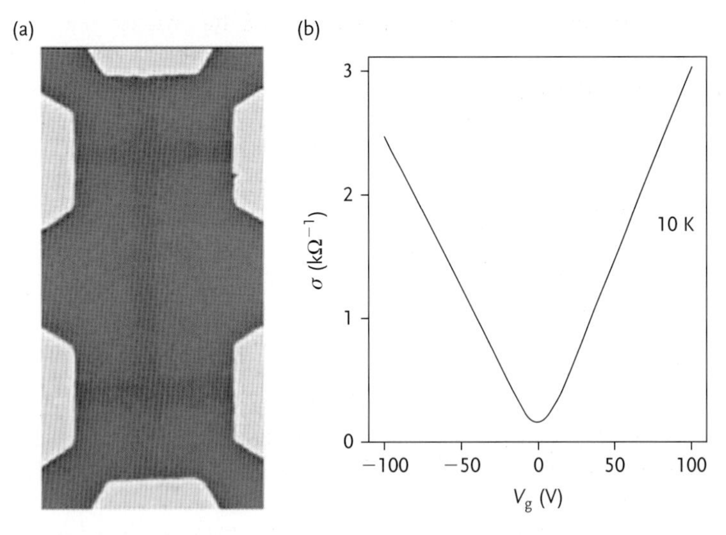

Figure 6.14 (a) The conductivity of a graphene layer [368] is measured in the Hall bar geometry (the center strip is about 0.2 μm wide). (b) The conductivity as a function of gate voltage V_g shows the characteristic V-shape and a pronounced minimum around the origin. The mobility in this device was calculated to be around 0.4 m^2/VS. For comparison, graphene membranes have shown mobilities up to 20 m^2/VS. Reprinted by permission from Macmillan Publishers Ltd: Nature [368], copyright 2005.

possess a very low density of structural defects, and it can often be assumed that scattering at structural defects is not a critical limiter. The presence of surface adsorbates and trapped charges at the interface to the substrate were, on the other hand, identified as dominant contributions to scattering in many cases. A localized charge, which is introduced via the substrate or an adsorbate will modify the position of the Fermi level with respect to the Dirac point and therefore create a localized so-called "charge puddle." The charge puddles are responsible for small fluctuations in the chemical potential and therefore impede charge carrier transport and act as scattering centers. The charge puddles can be observed directly with STM [48, 385, 386]; see also example in chapter 2.

The influence of the substrate can be eliminated by the fabrication of graphene membranes, or by the positioning of graphene layers across a trench in the substrate. Annealing of the membrane leads to the desorption of adsorbates and further improves performance [376]. The membrane systems have delivered to date the highest mobilities at low temperatures of $200 \cdot 10^3$ cm^2 V^{-1} s^{-1} for charge carrier densities of $5 \cdot 10^9$ cm^{-2}. The minimum, experimental conductivity σ at the Dirac point for a vanishing charge carrier density is measured to be close to $\sigma_{min} = 4e^2/h$, theoretical values are often given as $\sigma_{min} = 4e^2/h\pi$.

The advantages and challenges in the implementation of graphene-based electronics emerge directly from our understanding of the electronic structure. The ultrahigh mobility and the symmetry of the bands around the Dirac point are the most important properties of graphene and drive the development of graphene electronics. The challenges are the lack of even a small band gap, which is necessary to achieve a sufficient on-off ratio in current device architecture, and the limited ability to dope graphene and preserve the Dirac cone at the same time. In the last two years considerable progress has been made in what was in the beginning considered a severe limitation, namely the fabrication of large area graphene sheets. A substantial number of approaches have been explored to achieve doping and band-gap opening in graphene, although it remains to be seen which method is suitable for incorporation in carbon-based electronic devices and circuits. However, due to the very rapid development of research in graphene it is very likely that by the time this books is published new methods for the modulation of graphene's electronic properties, and the synthesis of high quality graphene will have been developed.

Doping is achieved in common semiconductor materials by substituting a lattice atom with a dopant atom lacking an electron (hole), or providing an additional electron. The dopant atom introduces additional states in the band gap of the semiconductor, and the Fermi level moves from the midgap position for the intrinsic material towards the valence band minimum for an n-type dopant (donor), or toward the conduction band maximum for a p-type dopant (acceptor). The controlled modulation of the position of the Fermi energy in the gap is the prerequisite to control charge transport, and build a multitude of electronic device structures. However, if a carbon atom in the honeycomb lattice of graphene is substituted, for example by a threefold coordinated B-atom, this trivalent B-atom introduces a hole and disrupts the delocalized π-bonded electron system. As we have seen earlier, the band structure of graphene relies on the symmetry of the honeycomb lattice, and it will be modified by the introduction of even small amounts of dopant atoms within the lattice. The shape of the bands around the Dirac point can be modified through the introduction of non-sp^2 carbon atoms, and at the same time these structural defects will increase scattering, and thus limit charge carrier mobility.

The doping of graphene is clearly a challenge, and the major lines of inquiry are currently *surface transfer doping*, where charge carriers are introduced through adsorbate at the surface or the interface to the substrate, *chemical doping* via the covalent bonding of molecules to graphene, and *electrostatic doping*, which uses an electric field to move the Fermi level away from the Dirac point. The latter method is also used to break the symmetry in bi-layer graphene and introduce a band gap (see next paragraph).

The in-plane doping of graphene is comparable to the attachment of covalently bonded hydrogen or other molecules, which can donate charge carriers in a controlled manner to the graphene sheet. A high degree of control over dopant concentration and spatial distribution can be achieved in this way, but the covalent bonding to graphene leads to a breaking of in-plane bonds, and thus the introduction of a large concentration of scattering centers. The detrimental effect of covalent attachment on charge carrier mobility is already observed for relatively low concentrations of only a few percent. Loss of the integrity of the honeycomb lattice and breaking of the in-plane covalent bonds can be avoided by "indirect" charge manipulation through the adsorption of electronegative or electropositive materials such as organic molecules or metal deposits. The doping of a graphene layer can now be modified by injecting charge carriers from the relatively weakly bonded adsorbate into graphene's valence or conduction band. Metals such as K, Au, and some commonly used contact materials act as electron donors and produce n-doped graphene layers [387–391].

Current device architecture requires a relatively large on-off ratio, the ratio between currents flowing through the device in the on and off state, and this can easily be realized if the active material has a band gap. For comparison, the on-off ratio in current CMOS (complementary metal oxide semiconductor) technology is around 10^4. In particular, the band gap guarantees a very low off-current since it limits charge carrier transport. It is possible to design device architecture where a band gap is not required, albeit this necessitates a substantial shift in how computing is done, and thus a considerable investment and reassessment of current and highly successful strategies.

We have seen that doping requires the controlled introduction of a charge without modification of the band structure and lattice periodicity, but a band gap can only be introduced if we modify the underlying symmetry of the lattice and thus the band structure. By itself this seems to contradict our previous requirement to retain the Dirac cone while modulating only selected aspects of graphene's electronic structure. However, several strategies such as strain engineering, symmetry breaking through substrate or adsorbate interaction, electrostatic gates for bilayer graphene and quantum confinement have been proposed, and appear to be promising routes to open a band gap.

In contrast to a single layer of graphene, the unit cell of bilayer graphene contains four atoms since due to the Bernal stacking the A and B atoms in first and second layer are not equivalent. The bilayer band structure therefore has two bonding and two antibonding π-bands. Breaking the inversion symmetry between the two layers now leads to a deformation of the band closer to E_F and an opening of a small band gap [380, 383, 392, 393]. The magnitude of this gap can reach 200–250 meV, which is significantly larger than the thermal energy at room temperature. This has been achieved in a reliable and reproducible manner through the application of a gate voltage with a dual-gate construction within a field effect transistor set-up. Zhang *et al.* [381] have recently demonstrated that selective doping, and tuning of the band gap can be achieved simultaneously in the dual-gate geometry by manipulation of the relative gate voltages for top and bottom electrodes. The addition of a chemical dopant to the surface of the bi-layer stack has a similar effect: the impact of the deposition of K was observed by Bostwick *et al.* and the symmetry is in this case broken through the donation of electrons from K-surface clusters [377, 394–397].

6.4.2
Quantum Confinement and Carbon Nanoribbons

Band-gap opening via electrostatic doping, however, is not possible for single layer graphene and the introduction of quantum confinement appears currently to be one of the most promising strategies. We have already discussed quantum confinement in semiconductor nanowires, carbon nanotubes, metal clusters and quantum dots, and can directly transfer these ideas to graphene. Sometimes graphene nanoribbons are even described as "flat nanotubes" where the short dimension in the nanoribbon is the direction of confinement and corresponds to the nanotube diameter. Graphene nanoribbons have been the subject of many theoretical studies, which predict an increasing band gap as a function of nanoribbon dimensions as expected for a confined system in the nanometer size regime. In contrast to the tubular nanotubes where the direction of confinement closes in on itself, we must now take into account the shape of the nanoribbon edges, which can adopt an armchair or zigzag configuration, and can be saturated or dangling bonds. This 1D boundary can introduce additional electronic states, and more importantly lead to losses by scattering, and is treated like the surface of the 2D nanoribbon. Several theoretical studies [398–400] have focused in recent years on band structure control through nanoribbon dimensions and tailored edge structures, although the experimental realization of nanoribbons has not been satisfactory. In particular large-scale production of well-defined nanoribbons is not available to date, but it is possible to fabricate small numbers of carbon nanoribbons with well-defined dimensions to test the theoretical predictions, and the results so far have been promising.

In a recent experiment Han *et al.* [401] and Chen *et al.* [398] were able to measure the band gap in graphene nanoribbons as a function of ribbon diameter. Han's results are summarized in Figure 6.15 and illustrate clearly the increase in band gap with decreasing nanoribbon width. The band gap begins to open for ribbon widths around 50–60 nm and reaches a value of around 120 meV for a 10–12 nm ribbon. The measurement of the band gap is achieved by incorporating the nanoribbon into a device structure where the two ends of the ribbon are connected to a graphene source and drain electrode, and a gate contact is

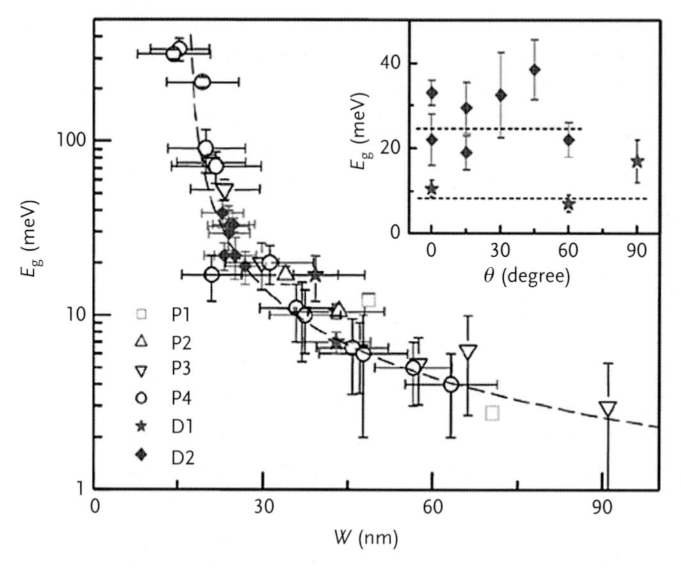

Figure 6.15 Relation between carbon nanoribbon width (W) and band gap (E_g) [401]. The nanoribbons were etched from a graphene sheet spanning the space between several metal leads, and six different devices are included in the figure. The inset shows that the band gap depends only weakly (if at all) on the relative angle of the graphene lattices in the ribbons. Reprinted figure with permission from [401].

positioned on top of the ribbon. The modulation of the source–drain voltage shifts the relative position of the valence and conduction band maxima, it introduces a chemical potential difference between the two ends of the ribbon, which drives a current through the structure. The gate voltage adjusts the position of the gap with respect to the source–drain voltage; it shifts the potential energy within the nanoribbon. This set-up is also used to fabricate single electron transistors, which are discussed in detail in Section 7.1. The conductance G of this mesoscopic structure is now measured as a function of the source–drain and gate voltage and the band gap of the nanoribbon can be measured. The conductance for nanoribbons, which are wider than about 60 nm and therefore do not possess a band gap, can be described by Ohm's law: $G = \sigma(W - W_0)/L$, where σ is the conductivity, and $W - W_0$ denotes the active width of a ribbon with length L. For ribbons with a band gap the conductance has to be described by a mesoscopic description, which takes into the account the limited number of "conduction channels" or states within the ribbon. The conductance in meso- and nanoscopic systems is discussed in detail for example in Datta's book [402] on "Electronic transport in mesoscopic systems."

Han *et al.* [401] made two additional observations, which are especially important to the development of nanostructures for electronic applications.

Firstly, the relative orientation of the graphene lattice with respect to the direction of the current does not appear to decisively influence the transport properties in his experiment. While a relatively small data-set is presented in their work due to the difficulties in measuring the crystallographic orientation of the nanoribbons, the effect is certainly small. This can be seen in comparison with carbon nanotubes, where the chirality of the tube directly determines its electronic structure. The origin of this discrepancy is not entirely clear, and it might be due to poorly defined edge structure in most graphene ribbons as Han *et al.* suggest in their conclusions. Secondly, the width of the nanoribbon is larger than the active region, which contributes to charge transport. This is incorporated in the calculation of G by using an effective width, which is reduced by W_0, an inactive edge region. The extension of the inactive reason increases with decreasing temperature, and is likely related to scattering processes, or the poor crystallinity of the edge region. This observation clearly shows the sensitivity of nanostructured materials to surface contributions, and control of the surface structure, and termination is critical to retain or even create functionality. Carbon nanoribbons are currently a very active area of research, and they offer an intriguing pathway to control electronic properties via the geometric structure. However, while it is theoretically understood how the geometry, edge structure and width of a nanoribbon affects its performance, the experimental challenges remain considerable.

6.5 Synthesis of Graphene

6.5.1 Exfoliation from Graphite

The first devices which were built to demonstrate the unique properties of graphene, used graphene sheets, which were extracted by exfoliation of highly oriented pyrolytic graphite (HOPG). The extraction included the repeated removal of a few layers of graphene at a time from a section of HOPG until only one single sheet of graphene is left, which adheres to a photoresist substrate, for example. This method is only feasible because of the large difference in bonding strength between inter-layer van der Waals forces, and intralayer covalent bonding, which allows the weakly bonded layers to be detached but leaves the 2D graphene sheets intact. This method has worked very well for the fabrication of proof-of-concept device structures [368, 403, 404], which were used to demonstrate many of the exciting physical and electronic properties of graphene. But HOPG, which is fabricated through high temperature pyrolysis of hydrocarbon precursors, consists of small crystallites, which are typically only a few micrometers in diameter, and the quality of HOPG is defined by their crystallographic misalignment. The mechanical exfoliation can therefore not be scaled, and device dimensions are limited by the HOPG's crystallography. Good quality single crystal graphite is rare and not a viable alternative.

6.5.2
Growth on Metal Substrates

The earliest work on graphene, which started a few decades before the recent surge in graphene research, was performed with graphene layers grown by pyrolysis of hydrocarbons on transition metal surfaces [405–411]. The layers often extend over relatively large areas, and therefore offer a route to the fabrication of extended graphene layers. It has recently been shown that very large and homogeneous graphene layers [412] can be grown on Cu foils by a CVD process, which opens the way to the fabrication of graphene sheets large enough to be compatible with parallel processing of devices. Careful selection of growth temperatures and cooling protocols are required to avoid film delamination during cool-down, and ensure the preferential growth of large, continuous and predominantly single layer material.

Figure 6.16 summarizes the most frequently used substrate materials [407–415]. The hybridization of the graphene's π-band and metal d-band determines the strength of the interaction and the distance between the metal surface and the graphene layer. Strongly interacting substrates such as Ni are akin to a chemisorbed layer with a relatively short bond-length between graphene and metal surface, while physisorbed systems such as Cu or Pt and graphene possess a larger interlayer distance. It can be assumed that the distortion of the graphene's electronic structure is much more pronounced in chemisorbed systems, albeit a complete dataset is not yet available.

The growth of graphene layers on metal surfaces proceeds via the dehydrogenation of hydrocarbon precursors, which is catalyzed by the metal

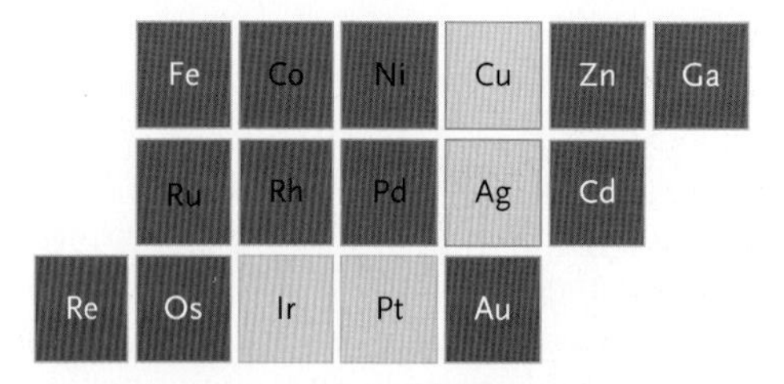

Figure 6.16 Section of the periodic table showing many of the commonly used substrate materials which promote graphene surface layer growth (black text), elements which do not promote graphene growth are marked with white letters. The weakly interacting metals are marked in light gray.

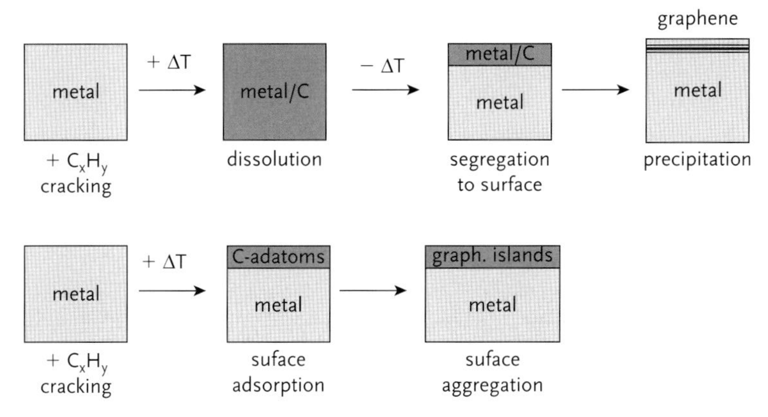

Figure 6.17 Illustration of two different mechanisms for the growth of graphene layers on metal substrates. The growth mode is determined mostly by the solubility of carbon in metal as a function of temperature. The illustration summarizes the results from Ref. [412].

itself and occurs at appreciable rates at elevated temperatures above about 500 °C. The next step in the growth process is controlled by the solubility of carbon in the metal bulk. The binary phase diagrams for Ni–C and Cu–C show that Ni possesses a relatively high solubility for carbon, while the carbon solubility of Cu is rather low. Li *et al.* [412] compared the growth mechanisms of graphene layers on these two metals, and were able to show in a rather elegant experiment, that these characteristic differences in carbon solubility lead to the emergence of two distinctly different growth modes, which are schematically shown in Figure 6.17. Ni and Cu do not form stable carbides, albeit a recent study indicates that Ni carbide might selectively form at lower temperatures as a surface compound [415]. In the case of Ni, which is representative of many transition metals, carbon is dissolved in the metal bulk and precipitates to the surface during the cool-down cycle. This reaction sequence is summarized in row 1 of Figure 6.17. The segregation of multiple layers can be avoided by using a rapid cool-down, which limits the mobility of carbon in the metal bulk and thus stops the growth of additional graphene layers. However, the graphene layer growth is a predominantly surface-driven process on Cu (row 2 in figure 6.17): since only minute amounts of carbon can be dissolved in the bulk, the nucleation and growth of graphene layers occurs at the surface, and segregation of carbon does not play a significant role. The study of the kinetics of graphene layer growth, unraveling the role of defects and grain boundaries, and their impact on the quality of the graphene layer are an active area of research, and of prime importance for future graphene device manufacture.

6.5.3 Sublimation of Si – Graphene on SiC

The disadvantage of graphene growth on metal substrates is the presence of the highly conductive substrate, and the interaction between graphene and many metals, which often distorts the Dirac cone. The metal substrate therefore has to be removed prior to device fabrication and the study of graphene's properties. An alternative is the synthesis of graphene on an insulating substrate such as the wide-band gap semiconductor SiC. This is likewise a relatively old method, and one of the first detailed studies on the mechanism of surface graphitization in SiC [416] was published in 1975. In this study the formation of a single layer graphene was recognized but the significance of this observation was not seen at the time. When SiC is heated to temperatures in excess of about 1000 °C the solid will begin to decompose by sublimation of Si from the surface. The now carbon-rich surface layer rearranges and a carbon interfacial layer is formed, followed by the growth of a high quality graphene layer. The growth rate of the carbon layer can be controlled by the temperature.

6H-SiC (the most frequently used SiC polytype) presents two different surfaces: the Si-terminated (0001)-SiC surface, and on the opposite side of the crystal, the C-terminated $(000\bar{1})$ SiC surface. On the Si-terminated surface, the formation of a single graphene sheet is followed by the growth of AB (Bernal)-stacked graphite, while on the C-terminated surface the graphene sheets form stacks of misaligned layers, which are electronically decoupled [325, 417–421]. The development of the surface structure during decomposition has been observed in considerable detail with spectroscopic and diffraction methods such as low electron energy diffraction (LEED). One of the main points of discussion is the structure of the intermediate layer between the still pristine SiC and the graphene honeycomb lattice. The structure of this interface is important for the continued growth of a graphene layer and it can influence its electronic structure. Several models have been considered over the years, including a Si-rich interface, an abrupt interface, and a covalently bonded carbon layer, which is currently the favored model. The structure of the interface is controlled by local strain due to the mismatch between the SiC reconstruction and the honeycomb lattice, and the high energy cost to the formation of a commensurate graphene overlayer. On the (000–1) face the interfacial layer is presumably a twisted carbon layer, which consequently leads to the formation of the misaligned graphene stack. These stacked layers are indeed electronically decoupled, which has been shown with angle resolved photoelectron spectroscopy [384]: several Dirac cones can be seen within the E–k diagram, and they are shifted with respect to each other due to the relative rotation of the graphene sheets. Qi *et al.* [419] have suggested that the interfacial layer on the (0001) surface contains defects with 5- and 7-membered rings to accommodate interfacial strain. The graphene layer on top of to the interfacial layer then exhibits a perfect honeycomb lattice.

6.6 Closing Remarks

In light of the challenges encountered in adapting graphene to the requirements of electronic device architecture, it is possible that this 2D material will not find its way to large scale applications, even though it allowed many aspects of the solid state physics of Dirac particles to be explored and understood for the first time. However, the class of 2D crystals is much larger and includes the highly ionic hexagonal BN, which has a large band gap, MoS_2, which has recently been used successfully in a field effect transistor (FET) prototype, and many other layered materials. The large anisotropy in bond strength, which is the prerequisite for the extraction of single, stable layers, is by no means unique to graphite and offers a wide field for exploration in the future.

7
A Few Applications of Inorganic Nanostructures

7.1
Single Electron Transistor

The applications for nanomaterials are numerous, and it would easily be possible to fill an entire book even if one chose to describe only the most important ones. However, a few applications have been built with many different nanomaterials, and exhibit an astonishing versatility. The single electron transistor (SET) is an example of one of these applications, and is at the same time a tool for probing electronic levels in a wide range of nanomaterials.

The single electron transistor (SET) is one of the most interesting devices, and it derives its functionality from the isolation of a charge on a nanoscale unit. An SET is a three terminal device and acts like a switch, where the switching action is driven by the transport of a single charge. The behavior of the SET is dictated by quantum mechanics and the specific properties of the nanoscale structure, which is the heart of the SET set-up. The SET function is determined by the size of the nanoscale structure, and a wide range of nanoscale units can be used including metal clusters, short segments of semiconductor nanowires, quantum dots (QD), graphene ribbons, or local doping on an Si surface [100, 180, 424].

The general layout of an SET is sketched in Figure 7.1: it consists of a nanoscale unit, which is connected through a tunneling barrier to the two electrodes, source and drain. The gate electrode is isolated from the nanoscale unit by an oxide layer and can be constructed through oxidation of the substrate area on which the SET structure is placed. The goal of this set-up is to guarantee the transport of single electrons onto and off the nanoscale unit by control of the electrode and gate voltages. In a conductor the current is perceived as continuous and the discreteness of the electron charge is only reflected, for example, in the Shot noise, which can be directly derived from the discrete nature of the elementary charge. If we want to control the transport of a single discrete charge it is necessary, firstly, to introduce a tunneling barrier, which localizes the charge on either

Inorganic Nanostructures: Properties and Characterization, First Edition. Reinke, P.

Published 2012 by WILEY-VCH Verlag GmbH & Co. KGaA

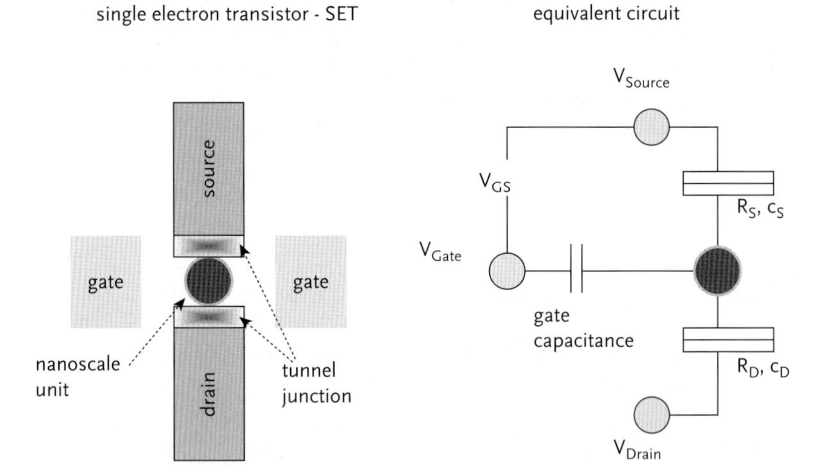

Figure 7.1 Schematic depiction of a single electron transistor (SET) and the corresponding equivalent circuit. The nanoscale unit is at the center of the SET and connected capacitively to the gate electrode, and coupled to the drain and source electrode by tunneling junctions with a characteristic capacitance and resistance.

side of the barrier, and secondly, to add a nanoscale unit, which is positioned within the barrier and captures the charge during its motion through the barrier. The nanoscale unit is connected to source and drain by two tunneling barriers, which can be described by their capacitance and resistance in the SET equivalent circuit. The gate is coupled capacitively and used to move charges within the SET. The charge transport through the nanoscale unit (or island) is controlled by the tunneling barriers and the amount of charge collected on the island. Once a charge tunnels onto the nanoscale unit, the Coulomb repulsion will prevent the addition of a second charge. The addition of more charges can be achieved if the charging energy is overcome by the potential difference between source and drain. The transport of single charges through an SET can only be done in a controlled manner if the charging energy, the energy required to move a charge onto the nanoscale unit, is significantly larger than k_BT, where k_B is the Boltzman constant.

In the simple case of a spherical condensor, which represents a spherical nanoparticle quite well, the charging energy U is given by:

$$U = e^2/2C \tag{7.1}$$

where e is the elementary charge; C is the capacitance of the spherical condensor of radius R and is given by:

$$C = 4\pi\varepsilon_0 R \tag{7.2}$$

where ε_0 is the dielectric constant. However, the charging energy of nanoscale units is a function of their geometry and size and can require considerably more complex description for the capacitance of the unit.

The charging energy U required to add a single charge is called the Coulomb blockade: if the current is plotted as a function of the source–drain voltage V_{SD}, no current will flow until V_{SD} is large enough to overcome the charging energy and an electron can move on and of the nanoscale unit. The Coulomb blockade is characterized by a plateau in the I–V characteristics. The stepwise addition of charge, the Coulomb staircase, can be seen very well in Figure 7.2, which shows the I–V characteristics for a spherical Au particle within an SET structure. In this experiment the electrodes (and bottom gate electrode) were fabricated using conventional lithography and Au was then deposited onto this electrode structure. Au forms small clusters with diameters of several nanometers, and those which are positioned correctly within the electrode gap will perform as nanoscale units in an SET. The Coulomb blockade can be seen clearly around $V_{SD} = 0$, and then the addition and removal of charge occurs in discrete steps, whose width is determined by U.

The effect of the gate voltage is seen clearly in the current–voltage characteristics shown in Figure 7.2: the Coulomb blockade until the first electron is moved onto the nanoscale unit is centered at $V_{SD} = 0$ for $V_G = 0$. When a gate voltage is applied the Coulomb blockade and staircase move in response to the applied voltage since the potential energy of the nanoscale unit is changed. The modulation of the current through the SET as a function V_{SD} and V_G is captured in so-called diamond diagrams, which are included in Figure 7.2. For the current throughout the SET we are usually interested in the variation of the current, and the position of the current steps as a function of external voltage. Instead of the absolute current the conductance $G = I/V$ (unit: Siemens = Ohm^{-1}) is displayed, and the step in current when an additional charge is added appears now as a "bright" (or dark) line within the diamond diagram and is easily distinguished from the plateau of the Coulomb blockade and staircase. The relative size of the diamonds is directly related to the stability of the system as a function of the number of electrons placed on the nanoscale unit. Kouwenhoven *et al.* [426] have shown this for QDs (Figure 7.3), and Bjork *et al.* [100] for QDs on nanowires (Figure 7.4), and the larger diamonds reflect the presence of an exceptionally stable charge state. This can be related to the filling of subbands within an electronic shell defined by the size and shape of the QD, where a completely filled shell is particularly stable. The magnitude of the charging energy is therefore also a function of the stability of the charge-state of the QD.

The regime of the SET operation described here is controlled by the capacitance of the nanoscale unit, and the capacitance and resistance of

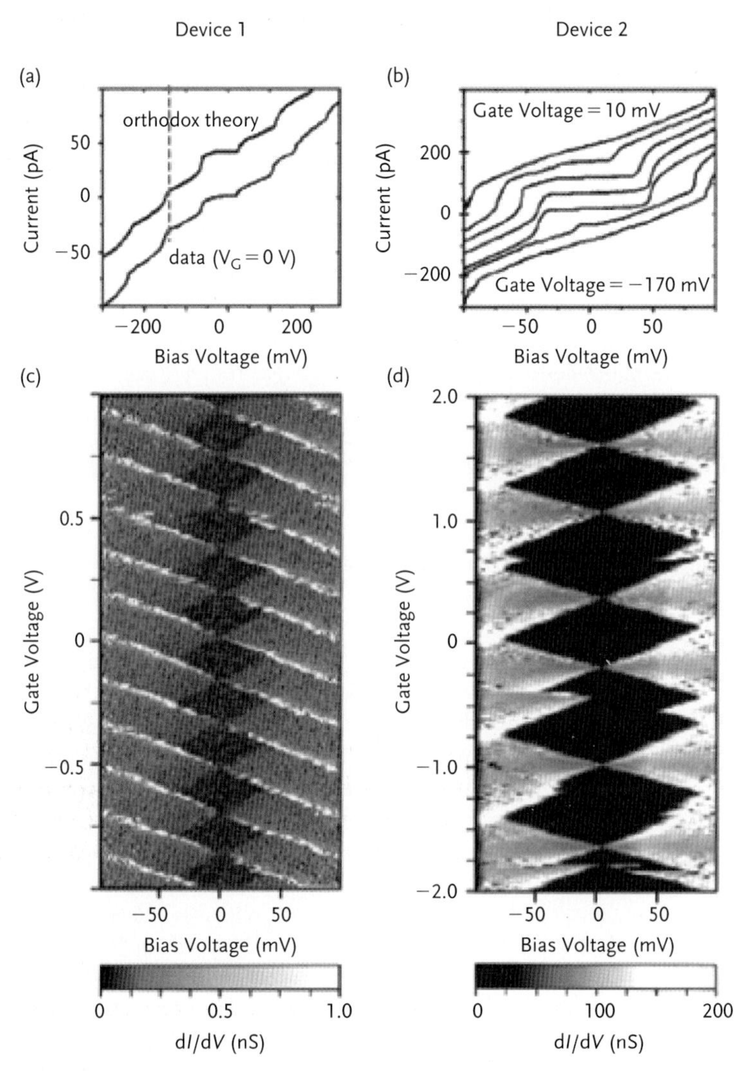

Figure 7.2 Examples for a diamond diagram from [180]. The SET consists of a Au–metal cluster, which is positioned between the source and drain electrodes. The substrate itself functions as a gate and is isolated from the device by an oxide layer, which provides the capacitive coupling. Device 1 and device 2 are distinguished by the position and size of the Au–metal cluster in the electrode gap. The Coulomb staircase graph for device 1 shows a comparison between the experimentally measured staircase and theory, and the graph for device 2 illustrates the shift of the Coulomb staircase with the gate voltage V_G. Reprinted with permission from Appl. Phys. Lett. [180]. Copyright 2004, American Institute of Physics.

the tunneling contacts themselves. We assumed that the transport onto the nanoscale unit is only limited by the amount of charge on the unit and therefore the coulombic interaction. However, as we decrease the size of the nanoscale unit, this assumption begins to break down, and the discrete

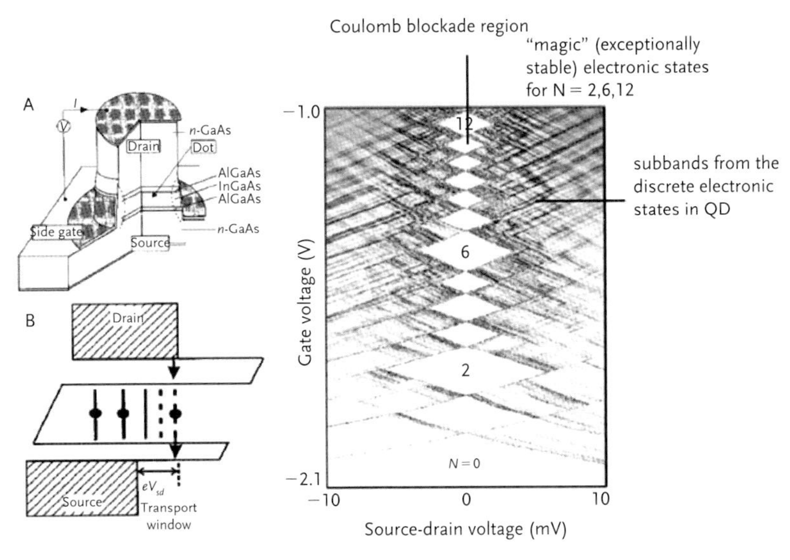

Figure 7.3 Diamond diagram of a semiconductor quantum dot [426], which is embedded in a semiconductor matrix. The junctions and tunneling barriers are defined by the respective band gaps of the semiconductors around the circular QD; the structure is schematically depicted in (A). This diagram shows the Coulomb blockade around $V_{SD} = 0$, the modulation of the magnitude of the Coulomb blockade as a function of the stability of the charge states on the QD, and the discrete levels/sub-bands on the QD, which emerge as a consequence of quantum confinement. The transport of charge through the QD can only be achieved if an electron from the source (or drain) can tunnel into an electronic state/sub-band on the QD. This situation is shown schematically in (B). From Science [426]. Reprinted with permission from AAAS.

energy levels within the nanoscale unit will define the transport of charge. This transition has been observed in numerous systems, and occurs if the spacing of the energy levels in the nanoscale unit becomes of the order of the charging energy. In the example shown in Figure 7.3 the discrete states, which are engaged in the resonant transport through the nanoscale unit, are seen as the narrowly spaced conductance maxima on either side of the inner diamond boundary. The transition is a function of the size of the nanoscale unit Bjork *et al.* [100] specifically address this correlation, and the results are shown in Figure 7.4. In this study the nanoscale unit of the SET is built into a semiconductor nanowire by switching the composition of the wire along the growth direction from small band gap InAs (0.3 eV) to wide band gap InP (1.4 eV), and back. The band gap and band offset between the InP and InAs regions provide the confinement potential, which defines the InAs QD. This geometry allows for easy modulation of the size of InAs quantum-dot region and 100-nm to 10-nm dots were investigated. Figure 7.4 shows the transition from the Coulomb repulsion controlled regime, which is characterized by regularly spaced peaks in the conductivity, to the regime where the energy levels in the QD dominate the current transport. The conductance spectrum now shows irregularly spaced peaks, whose position

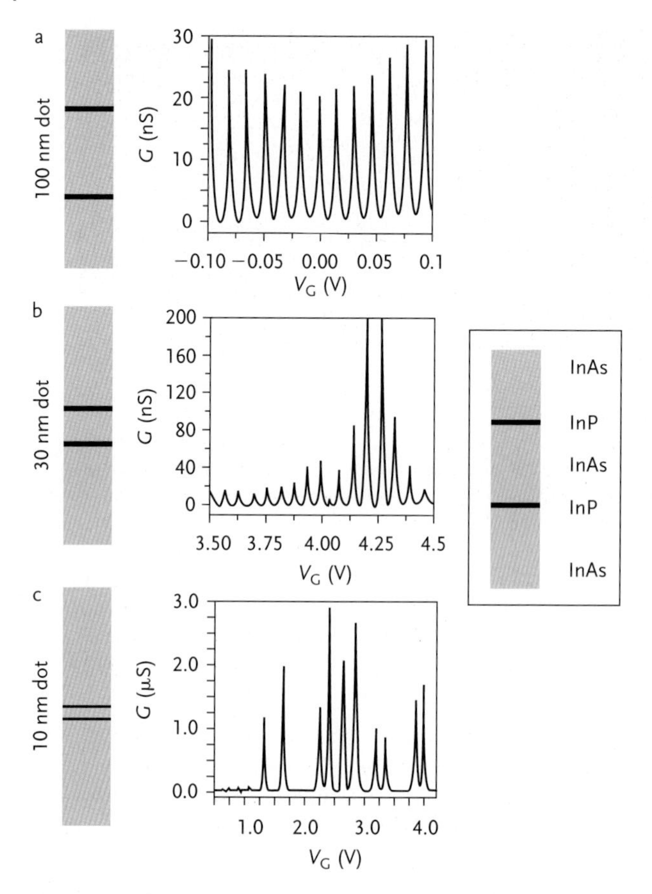

Figure 7.4 The observation of conductance G as a function of the size of the nanoscale unit shows the transition from the Coulomb blockade regime to the regime where quantum confinement and resonant tunneling control transport. This example is taken from [100] and the nanoscale unit is made from an InAs semiconductor wire, the tunneling barrier is defined by the introduction of InP segments. The longest InAs unit is 100 nm, and the even spacing of conductance maxima is due to Coulomb blockade, while for the 30 and 10 nm InAs units the maxima in G then reflect the sub-band structure on the nanoscale unit. Reprinted with permission from Nanoletters [100]. Copyright 2004 American Chemical Society.

is controlled by the resonant tunneling through the electronic states on the QD.

Studying the transport through a nanoscale unit in an SET configuration can also be used to investigate the energy levels within nanoscale structures. An excellent example of this application is the recent fabrication of what is probably the smallest SET currently available (see Figure 7.5): a Si(100) hydrogen-terminated surface is locally depleted of hydrogen using a scanning tunneling microscope (STM), and phosphorus as dopant atoms

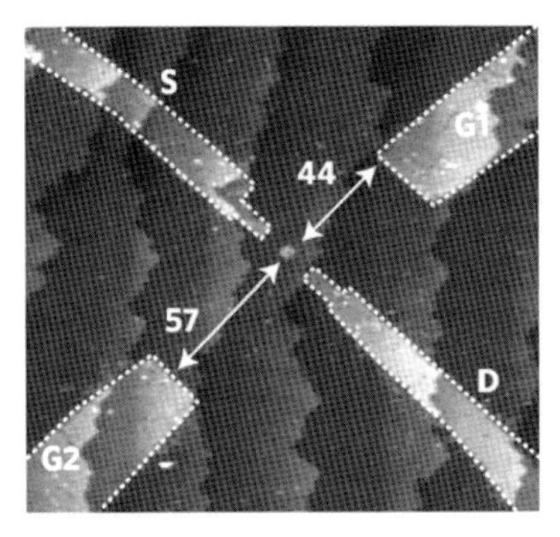

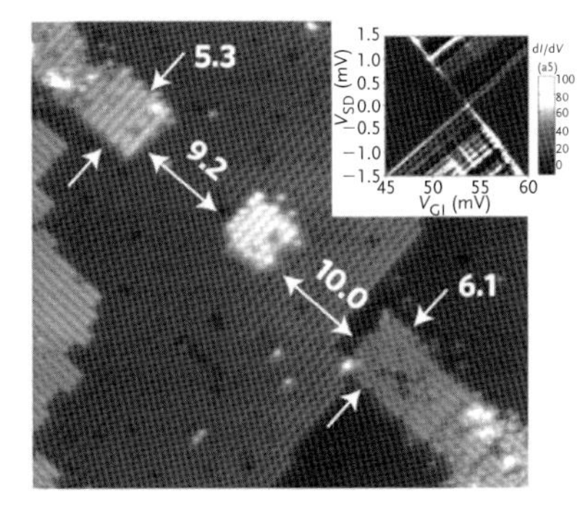

Figure 7.5 One of the smallest SET systems produced to date is shown here [424]. The nanoscale unit (QD), gate, source, and drain electrodes are "written" in the Si(100)-(2 × 1) surface by local removal of hydrogen with an STM. These sections look brighter in the STM images, and are in the next step saturated with P atoms as dopants. The tunneling gaps between the QD and the source and drain electrodes can be seen clearly, and the capacitive junction to the gate electrodes is created by adjusting the distance to the QD. The insert shows a section of the diamond diagram. Reprinted by permission from Macmillan Publishers Ltd: Nature Nanotech. [424], copyright 2010.

are attached on the hydrogen-free patch. This forms the nanoscale unit within the SET and the confinement is provided by local doping and the surrounding H-terminated surface. The leads are fabricated in a similar manner and the tunneling gap can clearly be identified in the STM, since it is a fully H-terminated section between the P-doped quantum dot-like patch and the leads. The corresponding diamond diagram is highly structured, and features a multitude of different energy levels, which are discussed in detail in the corresponding publication [424].

The operation of an SET cannot be fully comprehended without at least a basic understanding of nanoscale transport. When considering transport of charge through a metallic solid, we take into account the shape of the Fermi surface, the band structure, and density of states at the Fermi level. A continuum of states is assumed and the number of electrons and states available for conduction at the Fermi level is large. However, if charge transport through a molecule or nanoscale unit is considered, the discreteness of the energy states becomes a critical parameter. In each nanodevice we will encounter charge transport through an interface between a continuum of states (e.g., a contact pad) and a nanoscale unit with discrete energy levels. This is often described as mesoscopic transport; the size regime between macroscopic and nanoscale dimensions. The conductance through this interface is critical to the functionality of all nanoscale devices, including the SETs discussed above, molecular electronics devices and many more. The transport through the interface/junction is described by the conductance G, which in the Landauer formula is given as:

$$G = \frac{2e^2}{h} MT \tag{7.3}$$

where M is the number of *transverse modes* or *sub-bands* in the nanostructure which contribute to conduction within an energy interval, and T is the transmission coefficient through the nanoscale unit. Transport through the nanoscale unit can only be achieved if the Fermi level and conduction band of the macroscopic lead and contact lines up with energy states within the nanoscale unit. In this case T will increase rapidly and we observe resonant tunneling. A detailed and thorough discussion of many aspects of mesoscopic transport is given in S. Datta's book [402] on "Electronic Transport in Mesoscopic Systems." The resonant tunneling leads to the prominent lines of large conductance, which are seen in the diamond diagrams in addition to the Coulomb blockade steps and plateaux. Measuring the conductance throughout the nanoscale unit therefore allows its electronic structure to be identified.

7.2 Sensing with Graphene and Carbon Nanotubes

Sensing is not a single well-defined area of work, but includes a vast array of tasks, materials, electronic devices, and circuits. In the broadest sense of the word, sensing describes the transformation of a physical response to an event into an electrical signal, which can then be used to initiate an action or response. In remote sensing the physical response is triggered without physical contact and can be mitigated by electromagnetic radiation, pressure changes (e.g., wind), or radiation. Examples of remote sensing are the recording of atmospheric weather data, and the use of probes and satellites in the exploration of space. Sensors are indeed what the word already implies: an extension of our own senses to measure and analyze the world around us, and provide "artificial senses" as an interface with the physical world.

7.2.1 Sensors Everywhere

Our body is a complex package of sensors: tactile sensing with mechanoreceptors placed most densely in our fingers, optical sensing with the visual system made of a group of sensors adapted to different parts of the visible spectrum, auditory sensors for hearing, which is accomplished by the detection of subtle pressure changes, and chemical sensors, which are at the root of our sense of smell and taste. These are what we commonly refer to as our "five senses" and are the sensors we are most aware of in daily life. All of these sensors are built in such a way that their response to stimuli is transformed into an electric pulse, transmitted throughout the body via a highly interconnected nervous system, to the brain, which coordinates and most importantly, interprets the signal. The details of the human body's sensory system are far more complex, and this short paragraph is only an illustration of the general layout, which is indeed replicated in many artificial sensors.

An ideal sensor (Figure 7.6) will exhibit:

1. **Specificity/Selectivity** to the input of interest. A mechanical sensor should only be sensitive to mechanical force input and not to stray electromagnetic fields. An accelerometer on the other hand is sensitive to a change in the force acting on it rather then the magnitude of the force once a critical threshold is crossed. Specificity can be introduced through many different aspects of sensor design, and one example is functionalization of the sensor surface to enable reactions with a single type of molecule. Biological sensors often work this way and use surfaces with highly specific receptor molecules for detection.
2. **Reproducibility of response**. The same input should generate an identical output signal every single time the sensor is activated. The degree of reproducibility of a signal in response to identical stimuli also defines the reliability of the system. Ideally the signal transfer does not change over time, the sensor completely "forgets" the previous response, and the signal is reset to the reference level within an acceptable time interval.
3. **Independence**. A sensor and its electrical system interact only passively with the environment and do not change it in any way.
4. **Sensitivity** Small changes in input lead to large changes in output signal. This characteristic relationship between the magnitude of the input and output signals is captured in the transfer function. A linear transfer function with a constant sensitivity extending over many orders of magnitude would often be the ideal case, but can rarely be realized. Knowledge of the transfer function is therefore critical for sensor calibration and a quantitative analysis of the signal.

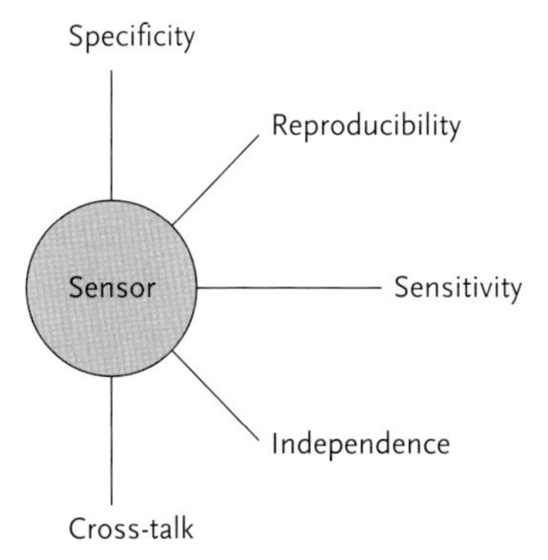

Figure 7.6 Schematic summary of properties which guide sensor design.

5. **Cross-talk**: Cross-talk is the interference with the sensor by triggering an output signal through an undesirable interaction on the input side, in signal conduction, transmission, or processing. The latter can often be avoided by modifying the design of the electronic circuit, however, signal cross-talk at the input can be more difficult to remove since it is usually related to the fundamental materials property used to elicit a sensor response. For example: a chemical sensor relies on the change of resistance in a sensor material, which is triggered by absorption of a gas such as CO, but can respond in a similar manner to other gases, thus losing its ability to detect CO in a reliable manner. In gas sensors the cross-talk can often not be eliminated but the level and transfer function of cross-talk with other gases is known, and can therefore be taken into account in the operation.

Sensing with nanostructures has several advantages such as a high surface area, high packing density, potential for miniaturization of devices and thus ease of integration, ability to build sensor modules, which unite a diverse set of sensors in a small space, and overall low power consumption. In this section we give some examples of the use of nanotubes and graphene, because these are well-suited to illustrate a wide range of sensor functions, but they are by no means the only nanoscale material currently studied as sensor materials. Both, carbon nanotubes and graphene are used as chemical sensors, functionalized for biological sensing, and tested as force and mechanical sensors. All these application make use of very specific materials properties, and surface reactions.

7.2.2
Chemical Sensors – Adsorption

One of the first experiments which demonstrated the feasibility of using carbon nanotubes as chemical and environmental sensors with an extremely high sensitivity was the report by Kong *et al.* in 2000 [428]. The carbon nanotube was incorporated in a field effect transistor (FET) and acted as channel of the device. The conductivity of the nanotube channel, and the resultant change in device characteristics were found to be highly sensitive to ppm (parts per million) levels of NO_2 and NH_3 and exhibited a very fast response time, albeit a much slower recovery time. Both gases are important in air quality and industrial pollution monitoring, and the ability to detect very small concentrations is highly coveted. This study and the large amount of research, which followed [429–436], illustrate the challenges in truly understanding the underlying mechanism of the sensitivity of the electrical response to the chemical modification of the nanotube surface. Carbon nanotubes are being studied for the detection of several other small gas molecules, and a summary list is included in Ref. [436].

The carbon nanotube is integrated in an electrical device structure, whose current–voltage and device characteristics are modulated by changes in the conductivity of nanotubes. The adsorption of molecules on the nanotubes surface leads to changes in the electrical properties of the tube via so-called chemical doping, which is charge transfer or band bending due to the introduction of a surface dipole. In general a semiconducting tube will therefore display a larger change in conductivity upon adsorption due to a significant increase electronic states in the gap, which directly correlates with conductivity. The absorption characteristics of a molecule, which includes the type of bonding (physisorption, chemisorption) and the adsorption site (defect, tube surface), are therefore critical to nanotube sensor function. One of the drawbacks in sensing by chemisorption of molecules is unfortunately very often a sizeable activation barrier to desorption and recovery can require substantial heating of the material.

The adsorption of a highly electronegative molecule such as NO_2 is assumed to lead to a charge transfer between the nanotube valence band and the NO_2 LUMO. The density functional theory (DFT) calculations by Peng *et al.* [432, 433] shown in Figure 7.7 illustrate this interaction and the modulation of the local charge density after adsorption of NO_2. However, the situation is more complex, and the decomposition of NO_2 at the surface into NO and NO_3, as well as adsorption at defect sites have been suggested as mechanisms leading to the observed conductivity change. However, the most recent work leads in a very different direction [435], namely that the adsorption of NO_2 has the most impact if it happens at the nanotube–electrode interface and modifies the Schottky barrier. This

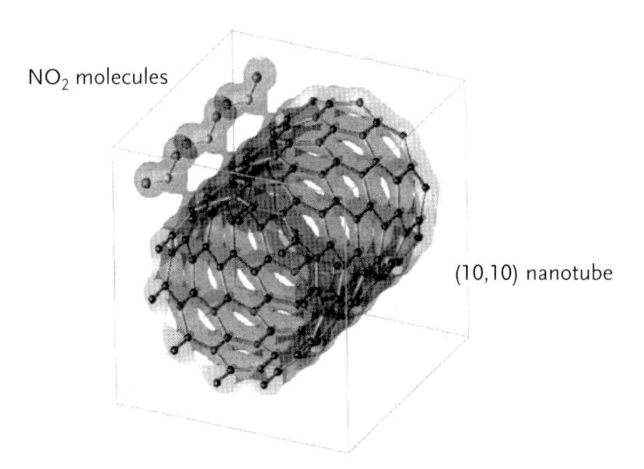

Figure 7.7 Calculation [432] of the modulation of the electron density along a (10,0) nanotube by the adsorption of 3 NO_2 molecules, each bound to a unit cell of the nanotube. The plot shows the valence electron charge density, and the tube is electron depleted. The bonding between tube and molecule is expressed in the continuity of the charge density contour including nanotube and NO_2. The figure is reprinted with permission from Nanotechnology [432] IOP Publishing Ltd. and the author.

long-standing discussion on the mechanisms, which is at the root of the nanotubes sensitivity to NO_2, illustrates the complexity of the process.

The interaction of small molecule adsorbates occurs in a similar manner on the graphene surface, and the differences between the adsorption sites are rather subtle. The most significant impact comes from the interaction of the charges within the molecules, and the π-system of the graphene layer. The modulation of the local graphene charge distribution by the adsorbates, however, has a very strong impact on the conductivity of the graphene layer due to the unique band structure of graphene, which directly correlates with the sensitivity. Schedin *et al.* [437] measured the effect of NO_2 adsorption on the conductivitiy of graphene sheets incorporated within a Hall bar geometry. NO_2 is the "molecule of choice" for many of these experiments due to its considerable electronegativity and thus a strong impact on conductivity is expected. The salient points of this experiment are summarized in Figure 7.8. The adsorption of NO_2 leads to a decrease in the Hall resistivity due to the injection of holes in graphene and a shift of the E_F from the Dirac point to the valence band, the desorption of NO_2 has the opposite effect and E_F is moved back to the Dirac point. The resistance change for addition/removal of an electron was calibrated in this case by the variation of gate voltage in the device. A jump or step in the Hall resistivity is equivalent to the addition/removal of a charge unit, and can be uniquely associated with adsorption or desorption of NO_2. The discreteness of the steps shows that it is possible to detect the adsorption and desorption of single molecules, the variation in step heights is due to a multitude of adsorptions sites, and overlapping events. The unprecedented accuracy for chemical sensing is solely due to graphene's unique band structure. However, this experiment also illustrates the sensitivity of the performance of graphene-based devices to small changes in processing and environmental conditions.

7.2.3 Tethering and Other Interactions

Chemical sensing via the direct adsorption on nanotubes and graphene is highly effective but also relatively unspecific. Molecules with large differences in electronegativity such as NO_2 and NH_3, or those with large differences in adsorption and desorption characteristics can be distinguished but the chemical specificity is still limited. In order to extend the selectivity and include the detection of larger and more complex, even biological, molecules it is necessary to functionalize the nanotube or graphene and tether a sensing molecule to the surface. In this way it is possible to make full use of the nanotube's/graphene's ability to translate a chemical into an electrical signal while greatly expanding the range of applications [438, 439]. For graphene the tethering of a sensing molecule is facilitated by

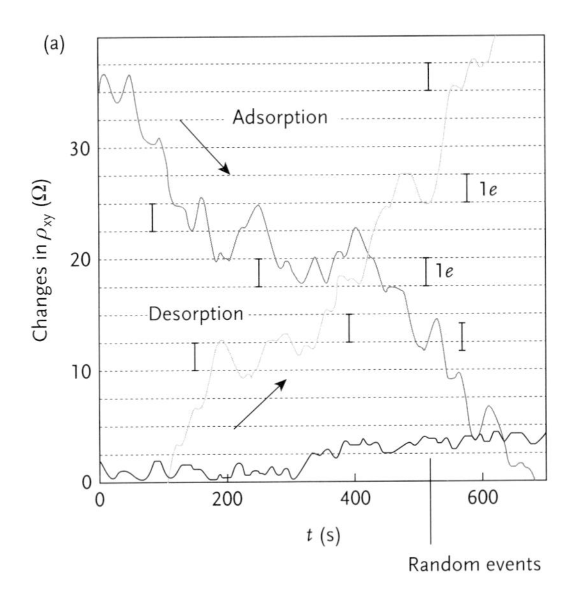

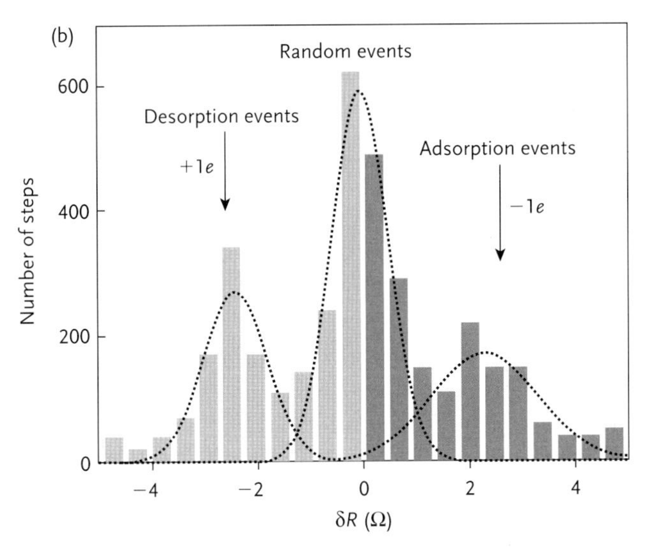

Figure 7.8 (a) Change in Hall resistivity of graphene (in a Hall-bar configuration) during the adsorption and desorption [437] of NO_2. The broken lines indicate the change in Hall resistivity, and each interval to the resistivity change corresponds to a single electron added or removed to the graphene sheet. The desorption curve was recorded during a mild heating of the graphene. The bottom curve is due to random events as they occur due to adsorption of desorption of environmental gases. (b) A histogram, which summarizes all the step-by-step changes in Hall resistivity recorded and summarized for a large number of events. The adsorption and desorption events of the highly electronegative NO_2 are characterized by a well-defined exchange of charge, and the single events can therefore be recorded. Reprinted by permission from Macmillan Publishers Ltd: Nature Materials [437], copyright 2007.

using graphite oxide (GO also called oxidized graphene), which is made by immersing graphite in a strongly oxidizing agent [440]. The introduction of carbonyl, hydroxide, carboxylic acid, and other oxygen-containing groups leads to a partial, local destruction of the graphene honeycomb lattice, but the conductivity modification of the layer is apparently still sufficient to be used for sensing applications.

Very recently the potential of graphene and nanotubes as interface between biological entities and electrical circuits has been recognized, and more and more novel nano–bio interfaces are described in the literature [441–443] (this is by no means a representative selection of the literature – only a few interesting publications to get started). One of these studies uses the mechanical deformation [441] of graphene, which is placed on a living cell, to study cell contraction and expansion. The cells expand or contract in response to stimuli and the resultant deformation of the graphene sheet leads to specific modulation of the graphene sheets resistance, which is then recorded. Another recent publication [439] reports the tethering of bacterial cells to chemically modified graphene sheets, and thus the creation of large cell ensembles which are immobilized on "detection" system namely the graphene sheet. The fabrication of highly selective graphene and nanotube based sensors is a wide and new field of research, and many new developments in the fabrication of nano–bio interfaces can be expected in the near future.

7.3 Quantum Dots, Rods, and Nanotubes in Photovoltaics

The use of fossil fuels to satisfy our ever increasing demands for energy is limited by a diminishing supply, and growing concerns about the environmental impact from rising atmospheric CO_2 levels. We need to rethink our strategies to use and supply energy and make it readily available while at the same time minimizing the environmental load we are placing on already highly stressed and polluted air, water, and land. It is likely that we will have to develop several approaches, which use wind, solar, geothermal, hydro-electric power, fossil fuels, energy conservation, and others in parallel. It would be highly desirable to cover all our energy needs from renewable energy sources but at our present rate of consumption it appears unlikely that we will be able to meet this goal. Nuclear power, which was once seen as a seemingly endless supplier of energy, has raised considerable concerns, many of them are related to long term use and storage of highly radioactive by-products. The recent accident at the Fukushima power plant in Japan has also shaken confidence in the construction and safety of the reactors themselves. There is probably not one solution, but many small solutions and new developments, which will play a role in future.

7.3.1
Solar Cells – a Short Introduction

In this section we will focus on one aspect of the whole spectrum of alternative energies, namely photovoltaics, and illustrate the use of inorganic nanostructures to improve their efficiency [444, 445]. A solar cell, the fundamental building block of a photovoltaics module, is by itself a fairly simple device [446–448]. The illumination of a semiconductor leads to excitation of an electron across the band gap from the valence into the conduction band. In this way an electron–hole pair (exciton) is formed, and the electron thermalizes to the conduction band minimum. This process was already described in Chapter 5 in our short excursion to optical properties. Instead of a recombination, which would only lead to the emission of another photon, the exciton is separated, the electron is collected by the cathode, the hole by the anode. The system has to be asymmetric in order to drive a current and consequently build up a photovoltage. The electric field to separate the charges can be supplied either by the application of an external bias voltage, or by the introduction of contact potentials, pn-junctions and Schottky barriers. A Si solar cell is essentially an "inverted" pn-junction (diode), instead of light emission by recombination of excitons, a current is driven by the creation and separation of excitons.

The quantum efficiency of a solar cell is defined by the probability that an incoming photon of energy E will produce one electron, which is delivered to the circuit. Power conversion efficiency (often only called efficiency) η is defined as the power output from the solar cell (the work which can be done by the cell) given by the product of open circuit voltage, current density, and filling factor divided by the power delivered by the sunlight to the surface of the cell. The question is how the efficiency of a solar cell can be optimized, and how these ideas are translated into a cheap, versatile, and robust modular device. The economic aspect is indeed very important in the development of solar cells, since they are produced for a very large, global market and can only become a relevant contribution to our energy portfolio if their price is competitive. Figure 7.9 illustrates schematically the development of price and efficiency in solar cells [446, 447], although new record-breaking efficiencies are reported on a nearly weekly basis. For example, the costs of bulk Si solar cells, which are often called first generation photovoltaics, are by now nearly entirely controlled by the cost of the high-quality Si material. This cost has somewhat been mitigated by using polycrystalline material, or "dirty silicon" [449] but the costs remain still relatively high. Second generation solar cells are thin film solar cells, where the amount of raw material is reduced but processing and encapsulation of the cell becomes a considerable contribution to the overall cost. Their efficiency is often lower than that for bulk cells due to a reduced absorption volume in the active region of the thin film, but they can more easily be adapted to cover large areas. A large portion of solar cell research is

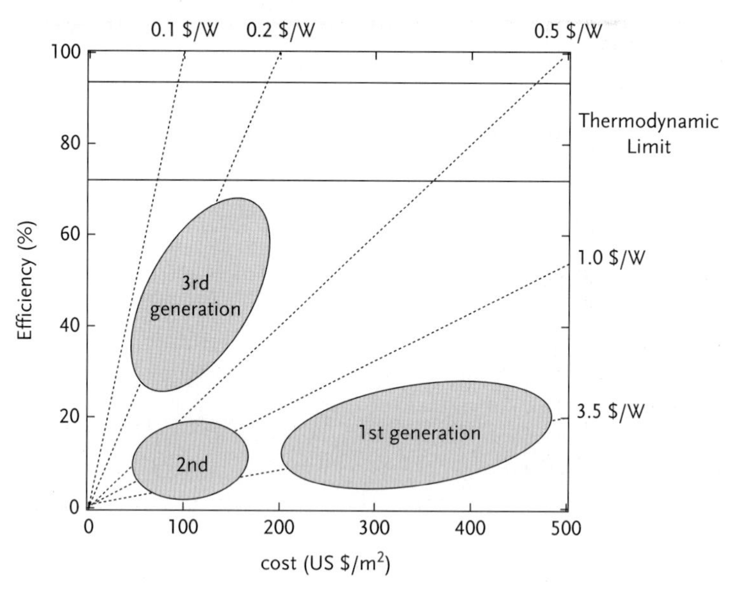

Figure 7.9 Schematic illustration of the relation between cost and efficiency of three generations of solar cells. The costs are based on the value of the US$ in 2003, and the efficiencies have for all types of solar cells been increased by several percent. However, despite these developments the overall picture shown here is still valid. The thermodynamic limit is given by the Carnot cycle, variations come from small changes in the overall assumptions. The Schottky–Queisser limit, which is the theoretical limit for a single junction cell is positioned at 31%. The figure is adapted from [447].

therefore currently directed toward the development of a third generation of solar cells, where the individual physical processes, which limit the quantum efficiency of the device, are optimized. In this section we will mention a few of the bottlenecks, which are encountered in the making the perfect, high yield, cheap solar cell. The underlying physical and chemical processes are complex, and optimization of one often comes at the cost of worsening the yield of another process. The challenge lies therefore in finding the optimum balance rather than perfect only one aspect of the solar cell functionality. The complexity of optimization of solar cell function leads to a vast array of solutions, which differ in many aspects of the solar cell architecture, structure, and materials.

The function of a solar cell can be subdivided roughly into three different segments: (i) light absorption, (ii) excitation and separation of electron and hole, and (iii) transport of the charge carriers to the electrodes. Inorganic nanomaterials as they are covered in this book, have been used in nearly every aspect of solar cell function: quantum dots, rods, and nanowires as variable band gap materials and broad-band absorber, QDs in hybrid cells to drive charge separation, carbon nanotubes as conductive material for highly efficient charge transport, and graphene to form highly transparent electrodes.

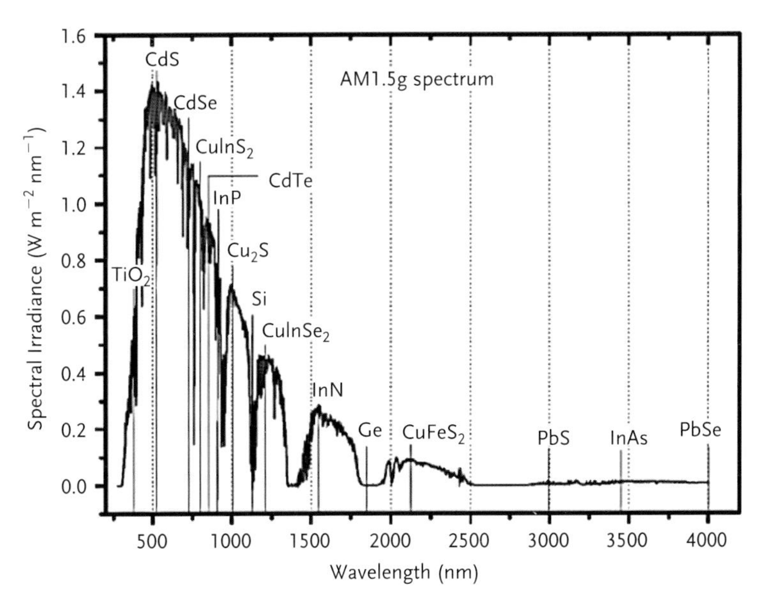

Figure 7.10 Solar spectrum [450] corrected for atmospheric adsorption (AM1.5). The bandgaps fo several common bulk semiconductor materials are indicated. From [450] Copyright Wiley-VCH Verlag GmbH & Co. KGaA. Reproduced with permission.

The advantages of nanomaterials lie in their versatility and adaptability of electronic and optical properties, and the ease of integration even in geometrically complex structures. A disadvantage is sometimes the large surface area, which can introduce interfacial traps and defects thus limiting exciton lifetime, hindering charge transport, and driving charge recombination.

7.3.2 Quantum Dots to Extend Absorption

The ideal solar cell should absorb across a wide range of the solar spectrum, ideally including the UV, visible, and near-IR regions. The solar spectrum can be approximated by the spectrum of a black body with a temperature of 5760 K and the AM1.5 solar spectrum [450] is shown in Figure 7.10; the band gaps of several semiconductor materials are included in the figure. The AM (atmospheric mass) 1.5 spectrum is corrected for absorption from the atmosphere, and the narrow lines are due to absorption by atmospheric gases. The absorption edge of the respective semiconductor defines the wavelength limit, longer wavelength/smaller energy photons cannot be absorbed since their energy is insufficient to excite an electron across the band gap. One solution to this challenge is the use of tandem solar cells, where layers of different semiconductor materials with decreasing band gap are stacked on top of each other. The use semiconductor QDs with a

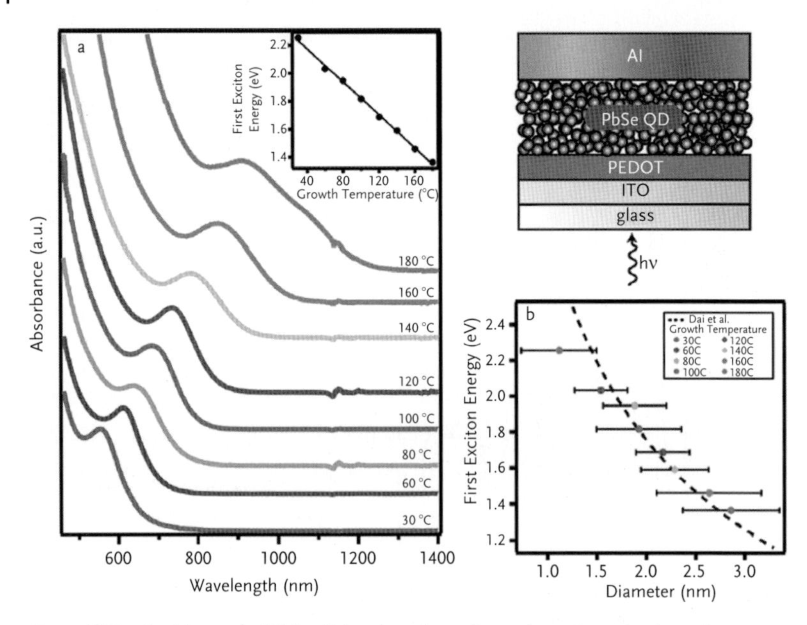

Figure 7.11 In this study [454] a Schottky solar cell was investigated, where the active layer is composed of ultrasmall PbSe quantum dots. (a) The absorption spectra [454] of PbSe QDs as a function of size; the correlation between QD diameter and bandgap [456] (first exciton energy) is shown in (b). The inset in (a) then relates the growth temperature to the first exciton energy and thus illustrates how control of the QD performance is achieved by adjusting processing condition. Reprinted with permission from ACS Nano [454]. Copyright 2011 American Chemical Society.

variation of size and therefore band gap follows the concept of a tandem cell, but avoids the complex processing required to produce stacked semiconductor layers [445, 450–455].

Figure 7.11 illustrates the change in absorption spectra as a function of PbSe QD diameter: the QD diameter increases with the growth temperature, and thus the band gap (first exciton energy) is reduced [454, 456]. The reduction in band gap leads to the shift of the absorption edge to longer wavelengths (this relation between band gap and absorption spectra was also discussed in Chapter 5). The PbSe QD mixture can now provide a high overall absorption of light across a wide range of the solar spectrum. In the set-up used in [454] a PbSe layer is placed between an Al electrode, which functions as the cathode, and a PEDOT[1] covered, transparent ITO (indium–tin oxide) anode, which provides an ohmic contact. This device architecture is often referred to as Schottky-junction architecture (see Figure 7.12), since the charge separation of the exciton occurs within the band-bending region in the semiconductor [457] (here the QD assembly) in contact with the metal

[1] Poly(3,4-ethylenedioxythiophene): conductive, optically transparent polymer.

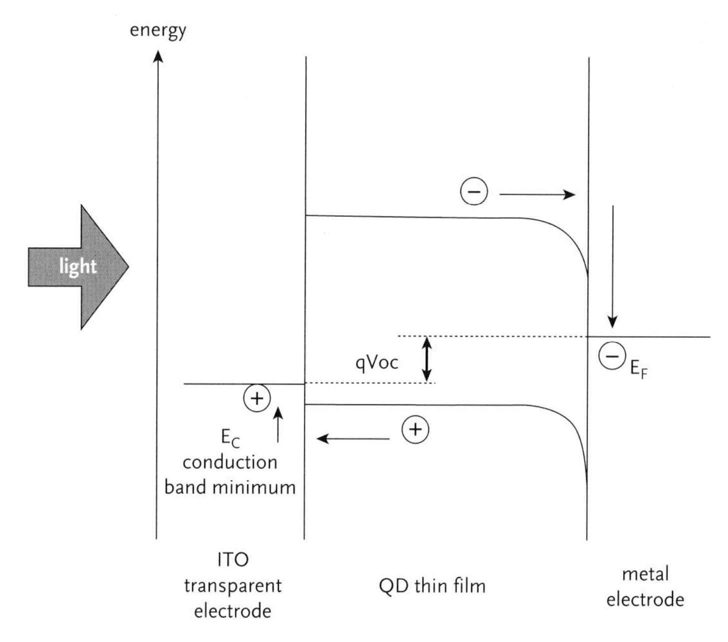

Figure 7.12 Schematic illustration of Schottky solar cell, with an Ohmic contact between the QD thin film and the ITO anode (Indium Tin oxide), and a Schottky barrier to the metal electrode. [The leakage currents (hole to metal electrode, electron to ITO) are not included.] The exciton creation is indicated at the center of the illustration, and qV_{oc} corresponds to the energetic difference between the Fermi levels (open circuit voltage), which drives charge separation. The arrows indicate the direction of transport for positive and negative charge.

electrode. Ideally one would like to have a very large volume for photon absorption and exciton generation, but on the other hand the exciton has to move to the region where charge separation occurs before it is destroyed by recombination. Therefore the distance to the charge separation region should not exceed the exciton mean free path as defined by its mobility and lifetime, otherwise recombination losses will diminish overall efficiency.

The mechanism charge and exciton transport depend on the nature of the material: for very small QDs charge transport occurs predominantly by hopping between the individual QDs, and it becomes less efficient if more hops are required to reach the charge separation region. One way to improve transport across an ensemble of nanostructures was suggested by Huynh *et al.* [458]: instead of spherical QDs they used nanorods of the same material. Absorption can occur anywhere in the nanorod, the absorption spectra are defined by quantum confinement due to the small rod diameter, but transport can occur easily along the long dimension of the wire thus limiting hopping across multiple interfaces. Another way to reduce the average distance an exciton must travel to the electrode is modification of the cell architecture, or morphology of the active region. However, relatively low overall mobilities remain a challenge in QD-based solar cells.

The Schottky QD solar cell is the simplest possible architecture and Kramer and Sargent [453] describe, in their review, the transition from the Schottky cell to more and more complex multilayer QD solar cell architectures. One particular arrangement, which has been quite successful, is the QD sensitized cell [452, 453, 459–461], which can be directly compared to dye-sensitized solar cells. In this architecture the QD is attached to a TiO_2 surface (e.g., a porous TiO_2 network produced by sintering), and the relative band energies at the interface between QD and TiO_2 facilitate exciton breakup. The electron is then injected into the TiO_2 while the hole is transported through an electrolyte to a metal anode. With this arrangement it is possible to use the wide range of band gaps offered by a mixture of QDs and at the same time mitigate the poor transport properties, large distance to electrodes, and recombination at the surface of QDs.

Until now we have concentrated on the active part of the solar cell, where the conversion from light to electricity occurs, but will now briefly turn our attention to the electrodes. In order to allow light to enter the active region of the solar cell at least one of the electrodes has to be transparent and conducting. This is accomplished by using ITO (Indium Tin Oxide), which is unfortunately a relatively expensive electrode material due to the costs of the metals In and Sn. The use of carbon nanotube composites [462–464] has recently been suggested to provide the unique combination of high optical transparency and favorable mechanical properties. The carbon nanotube film is embedded in a polymer and the high aspect ratio of the nanotubes guarantees a low percolation threshold and a larger number of charge transport paths through the network. However, the performance of the nanotube electrodes was limited by a relatively high sheet resistance, which has recently been overcome by using nanotube electrodes with a large percentage of metallic tubes.

There are many more examples of the successful use of inorganic nanostructures in photovoltaics, and many opportunities, which wait for their discovery. We hope that this short chapter can serve as a door to a much wider world of exploration in the future.

References

1 Bell, A.T. (2003) The impact of nanoscience on heterogeneous catalysis. *Science*, **299**, 1688.
2 Haruta, M. (1997) Size- and support dependency in the catalysis of gold. *Catal. Today*, **36**, 153.
3 Choudhary, T.V. and Goodman, D.W. (2005) Catalytically active gold: the role of cluster morphology. *Appl. Catal. A: Gen.*, **291**, 32.
4 Chen, F., Ramayya, E.B., Euaruksakul, C., Himpsel, F.J., Celler, G.K., Ding, B., Knezevic, I., and Lagally, M.G. (2010) Quantum confinement, surface roughness, and the conduction band structure of ultrathin silicon membranes. *ACS Nano*, **4**, 2466.
5 Hybertson, M.S. and Needels, M. (1993) First-principles analysis of electronic states in silicon nanoscale quantum wires. *Phys. Rev. B*, **48**, 4608.
6 Sanders, G.D., Stanton, C.J., and Chang, Y.C. (1993) Theory of transport in silicon quantum wires. *Phys. Rev. B*, **48**, 11067.
7 Yu, P.Y. and Cardona, M. (1996) *Fundamentals of Semiconductors – Physics and Materials Properties*, Springer-Verlag, Berlin, New York.
8 Crommie, M.F., Lutz, C.P., and Eigler, D.M. (1993) Imaging standing waves in a two-dimensional electron gas. *Nature*, **363**, 524.
9 Fiete, G.A. and Heller, E.J. (2003) Colloqium: theory of quantum corrals and quantum mirages. *Rev. Mod. Phys.*, **75**, 933.
10 Kittel, C. (2004) *Introduction to Solid State Physics*, Wiley, New York.
11 Zangwil, A. (1988) *Physics at Surfaces*, Cambridge University Press, Cambridge, UK.
12 Zaremba, E. and Kohn, W. (1977) Theory of helium adsorption on simple and noble-metal surfaces. *Phys. Rev. B*, **15**, 1769.
13 Oura, K., Lifshitz, V.G., Saranin, A.A., Zotov, A.V., and Katayama, M. (2003) *Surface Science: An Introduction*, Springer-Verlag, Berlin, New York.
14 Thiel, P.A. and Evans, J.W. (2000) Nucleation, growth, and relaxation of thin films: metal(100) homoepitaxial systems. *J. Phys. Chem. B*, **104**, 1663.
15 Kalff, M., Comsa, G., and Michely, T. (1998) How sensitive is epitaxial growth to adsorbates. *Phys. Rev. Lett.*, **81**, 1255.
16 Michely, T., Hohage, M., Bott, M., and Comsa, G. (1993) Inversion of growth speed anisotropy in two dimensions. *Phys. Rev. Lett.*, **70**, 3943.
17 Kalff, M., Smilauer, P., Comsa, G., Michely, T. (1999) No coarsening in Pt(111) homoepitaxy. *Surf. Sci.* **426**, L447.
18 Liu, H., Lin, Z., Zhigilei, L.V., and Reinke, P. (2008) Fractal structures in fullerene layers: simulation of the growth process. *J. Phys. Chem. C*, **112**, 4687.
19 Liu, H. and Reinke, P. (2006) C60 thin film growth on graphite: coexistence of spherical and fractal-dendritic islands. *J. Chem. Phys.*, **124**, 164707.

Inorganic Nanostructures: Properties and Characterization, First Edition. Reinke, P.

Published 2012 by WILEY-VCH Verlag GmbH & Co. KGaA

20 Ehrlich, G. and Hudda, F.G. (1966) Atomic view of surface self-diffusion: tungsten on tungsten. *J. Chem. Phys.*, **44**, 1039.

21 Schwoebel, R.L. and Shipsey, E.J. (1966) Step motion on crystal surfaces. *J. Appl. Phys.*, **37**, 3682.

22 Crommie, M.F., Lutz, C.P., and Eigler, D.M. (1993) Confinement of electrons to quantum corrals on a metal surface. *Science*, **262**, 218.

23 Manoharan, H.C., Lutz, C.P., and Eigler, D.M. (2000) Quantum mirages formed by coherent projection of electronic structure. *Nature*, **403**, 512.

24 Bennewitz, R., Crain, J.N., Kirakosian, A., Lin, J.-L., McChesney, J.L., Petrovykh, D.Y., and Himpsel, F.J. (2002) Atomic scale memory at a silicon surface. *Nanotechnology*, **13**, 499.

25 Fischbein, M.D. and Drndic, M. (2007) Sub-10 nm device fabrication in a transmission electron microscope. *Nano Lett.*, **7**, 1329.

26 Tsong, T.T. and Sweeney, J. (1979) Direct observation of the atomic structure of W{100} surfaces. *Solid State Commun.*, **30**, 767.

27 Binnig, G., Rohrer, H., Gerber, C., and Weibel, E. (1982) Tunneling through a controllable vacuum gap. *Appl. Phys. Lett.*, **40**, 178.

28 Binnig, G., Rohrer, H., Gerber, C., and Weibel, E. (1983) 7×7 reconstruction in Si(111) resolved in real space. *Phys. Rev. Lett.*, **50**, 120.

29 Bardeen, J. (1961) Tunneling from a many-particle point of view. *Phys. Rev. Lett.*, **6**, 57.

30 Tersoff, J. and Hamann, D.R. (1983) Theory and application for the scanning tunneling microscope. *Phys. Rev. Lett.*, **50**, 1998.

31 Tersoff, J. and Hamann, D.R. (1985) Theory of the scanning tunneling microscope. *Phys. Rev. B*, **31**, 805.

32 Barth, J.V., Brune, H., Ertl, G., and Behm, R.J. (1990) Scanning tunneling microscopy observations on the reconstructed Au(111) surface: atomic structure, long-range superstructure, rotational domains, and surface defects. *Phys. Rev. B*, **42**, 9307.

33 Hamers, R.J., Tromp, R.M., and Demuth, J.E. (1986) Surface electronic structure of Si(111)-(7×7) resolved in real space. *Phys. Rev. Lett.*, **56**, 1972.

34 Takayanagi, K., Tanishiro, Y., Takahashi, M., and Takahashi, S. (1985) Structural analysis of Si(111)-(7×7) by UHV-transmission electron diffraction and microscopy. *J. Vac. Sci. Technol. A*, **3**, 1502.

35 Tromp, R.M. (1989) Spectroscopy with the scanning tunneling microscope. *J. Phys.: Condens. Matter*, **1**, 10211.

36 Yin, W., Wolf, S., Ko, C., Ramanathan, S., and Reinke, P. (2010) Nanoscale probing of electronic band gap and topography of VO_2 thin film surfaces by scanning tunneling microscopy. *J. Appl. Phys.*, **109**, 024311.

37 Lang, N.D. (1986) Spectroscopy of single atoms in the scanning tunneling microscope. *Phys. Rev. B*, **34**, 5947.

38 Meyer, E., Hug, H.-J., and Bennewitz, R. (2004) *Scanning Probe Microscopy – The Lab on a Tip*, Springer-Verlag, Berlin, New York.

39 Giessibl, F.J. (2003) Advances in atomic force microscopy. *Rev. Mod. Phys.*, **75**, 949.

40 Bartels, L., Meyer, G., and Rieder, K.-H. (1997) Basic steps of lateral manipulation of single atoms and diatomic clusters with a scanning tunneling microscope tip. *Phys. Rev. Lett.*, **79**, 697.

41 Lyding, J.W., Shen, T.-C., Hubacek, J.S., Tucker, J.R., and Abeln, G.C. (1994) Nanoscale patterning and oxidation of H-passivated Si(100)-(2×1) surfaces with an ultrathigh vacuum scanning tunneling microscope. *Appl. Phys. Lett.*, **64**, 2010.

42 Shirai, Y., Osgood, A.J., Zhao, Y., Kelly, K.F., and Tour, J.N. (2005) Directional control in thermally driven single-molecule nanocars. *Nano Lett.*, **5**, 2330.

43 Huefner, S. (1996) *Photoelectron Spectroscopy: Principles and Applications*, Springer-Verlag, Berlin, New York.

44 Stoehr, J. (1992) *NEXAFS Spectroscopy*, Springer Verlag, Berlin, New York.

45 Bardyszewski, W. and Hedin, L. (1985) A new approach to the theory of photoemission from solids. *Phys. Scripta*, **32**, 439.

46 Yeh, J.J. and Lindau, I. (1985) Atomic sub-shell photoionization cross sections and asymmetry parameters:$1 \leq z \leq 103$ *At. Data Nucl. Data Tables* **32**, 1.

47 Doniach, S. and Sunjic, M. (1970) Many-electron singularity in x-ray photoemission and x-ray line spectra from metals. *J. Phys. C: Solid State Phys.*, **3**, 285.

48 Zhang, Y., Brar, V.W., Girit, C., Zettl, A., and Crommie, M.F. (2009) Origin of spatial charge inhomgeneity in graphene. *Nat. Phys.*, **5**, 722.

49 Hembacher, S., Giessibl, F.J., Mannhart, J., and Quate, C.F. (2003) Revealing the hidden atom in graphite by low-temperature atomic force microscopy. *Proc. Natl. Acad. Sci.*, **100**, 12539.

50 Boyen, H.-G., Ethirajan, A., Kaestle, G., Weigl, F., Ziemann, P., Schmid, G., Garnier, M.G., Buettner, M., and Oelhafen, P. (2005) Alloy formation of supported gold nanoparticles at their transition from clusters to solids: does size matter?. *Phys. Rev. Lett.*, **94**, 016804.

51 Lu, W. and Lieber, C.M. (2007) Nanoelectronics from the bottom up. *Nat. Mater.*, **6**, 841.

52 Samuelson, L., Thelander, C., Bjork, M.T., Borgstrom, M., Deppert, K., Dick, K. A., Hansen, A.E., Martensson, T., Panev, N., Persson, A.I., Seifert, W., Skold, N., Larsson, M.W., and Wallenberg, L.R. (2004) Semiconductor nanowires for 0D and 1D physics and applications. *Phys. E*, **25**, 313.

53 Wagner, R.S. and Ellis, W.C. (1964) Vapor–liquid–solid mechanism of single crystal growth. *Appl. Phys. Lett.*, **4**, 89.

54 Gutsche, C., Regolin, I., Blekker, K., Lysov, A., Prost, W., and Tegude, F.J. (2009) Controllable p-type doping of GaAs nanowires during vapor–liquid–solid growth. *J. Appl. Phys.*, **105**, 024305.

55 Kolasinski, K.W. (2006) Catalytic growth of nanowires: vapor–liquid–solid, vapor–solid–solid, solution–liquid–solid and solid–liquid–solid growth. *Curr. Opin. Solid State Mater. Sci.*, **10**, 182.

56 Tian, B., Zheng, X., Kempa, T.J., Fang, Y., Yu, N., Yu, G., Huang, J., and Lieber, C.M. (2007) Coaxial silicon nanowires as solar cells and nanoelectric power sources. *Nature*, **449**, 885.

57 Yang, P., Yan, R., and Fardy, M. (2010) Semiconductor nanowire: what's next?. *Nano Lett.*, **10**, 1529.

58 Ionescu, A.M. (2010) Nanowire transistors made easy. *Nat. Nanotechnol.*, **5**, 178.

59 Kelzenberg, M.D., Boettcher, S.W., Petykiewicz, J.A., Turner-Evans, D.B., Putnam, M.C., Warren, E.L., Spurgeon, J.M., Briggs, R.M., Lewis, N.S., and Atwater, H.A. (2010) Enhanced absorption and carrier collection in Si wire arrays for photovoltaic applications. *Nat. Mater.*, **9**, 239.

60 Law, M., Greene, L.E., Johnson, J.C., Saykally, R., and Yang, P. (2005) Nanowire dye-sensitized solar cells, *Nat. Mater.*, **4**, 455.

61 McAlpine, M.C., Ahmad, H., Wang, D., and Heath, J.R. (2007) Highly ordered nanowire arrays on plastic substrates for ultrasensitive flexible chemical sensors. *Nat. Mater.*, **6**, 379.

62 Minot, E.D., Kelkensberg, F., van Kouwen, M., van Dam, J.A., Kouwenhoven, L. P., Zwiller, V., Borgstroem, M.T., Wunnicke, O., Verheijen, M.A., and Bakkers, E.P.A.M. (2007) Single quantum dot nanowire LEDs. *Nano Lett.*, **7**, 367.

63 Givargizov, E.I. (1973) Periodic instability in whisker growth. *J. Cryst. Growth*, **20**, 217.

64 Persson, A.I., Larsson, M.W., Stenstroem, S., Ohlsson, B.J., Samuelson, L., and Wallenberg, L.R. (2004) Solid-phase diffusion mechanism for GaAs nanowire growth. *Nat. Mater.*, **3**, 677.

65 Gole, J.L., Stout, J.D., Rauch, W.L., and Wang, Z.L. (2000) Direct synthesis of silicon nanowires, silica nanospheres, and wire-like nanosphere agglomerates. *Appl. Phys. Lett.*, **76**, 2346.
66 Wang, N., Tang, Y.H., Zhang, Y.F., Lee, C.S., Bello, I., and Lee, S.T. (1999) Si nanowires grown from silicon oxide. *Chem. Phys. Lett.*, **299**, 237.
67 Wang, N., Tang, Y.H., Zhang, Y.F., Lee, C.S., and Lee, S.T. (1998) Nucleation and growth of Si nanowires from silicon oxide. *Phys. Rev. B*, **58**, R16024.
68 Wu, Y. and Yang, P. (2001) Direct observation of vapor–liquid–solid nanowire growth. *J. Am. Chem. Soc.*, **123**, 3165.
69 DeHoff, R.T. (1993) *Thermodynamics in Materials Science*, McGrawHill, New York.
70 Buffat, P. and Borel, J.-P. (1976) Size effect on the melting temperature of gold particles. *Phys. Rev. A*, **13**, 2287.
71 Breaux, G.A., Neal, C.M., Cao, B., and Jarrold, M.F. (2005) Melting, premelting, and structural transitions in size-selected aluminum clusters with around 55 atoms. *Phys. Rev. Lett.*, **94**, 173401.
72 Haberland, H., Hippler, T., Donges, J., Kostko, O., Schmidt, M., and von Issendorff, B. (2005) Melting of sodium clusters: where do the magic numbers come from?. *Phys. Rev. Lett.*, **94**, 035701.
73 Hendy, S.C. (2007) A thermodynamic model for the melting of supported metal nanoparticles. *Nanotechnology*, **18**, 175703.
74 Dick, K.A., Deppert, K., Mårtensson, T., Mandl, B., Samuelson, L., and Seifert, W. (2005) Failure of the vapor–liquid–solid mechanism in Au-assisted MOVPE growth of InAs nanowires. *Nano Lett.*, **5**, 761.
75 Wacaser, B.A., Reuter, M.C., Khayyat, M.M., Wen, C.-Y., Haight, R., Guha, S., and Ross, F.M. (2009) Growth system, structure, and doping of aluminum-seeded epitaxial silicon nanowires. *Nano Lett.*, **9**, 3296.
76 Wacaser, B.A., Dick, K.A., Johansson, J., Borgstroem, M.T., Deppert, K., and Samuelson, L. (2009) Preferential interface nucleation: an expansion of the VLS growth mechanism for nanowires. *Adv. Mater.*, **21**, 153.
77 Dick, K.A., Deppert, K., Larsson, M.W., Mårtensson, T., Seiffert, W., Wallenberg, L.R., and Samuelson, L. (2004) Synthesis of branched "nanotrees" by controlled seeding of multiple branching events. *Nat. Mater.*, **3**, 380.
78 Ohlsson, B.J., Bjoerk, M.T., Magnusson, M.H., Deppert, K., Samuelson, L., and Wallenberg, L.R. (2001) Size-, shape, and position-controlled GaAs nano-whiskers. *Appl. Phys. Lett.*, **79**, 3335.
79 Duffe, S., Groenhagen, N., Patryarcha, L., Sieben, B., Yin, C., von Issendorff, B., Moseler, M., and Hoevel, H. (2010) Penetration of thin C_{60} films by metal nanoparticles. *Nat. Nanotechnol.*, **5**, 335.
80 Kelly, A., Groves, G.W., and Kidd, P. (2000) *Crystallography and Crystal Defects*, 2nd edn, John Wiley and Sons, Chichester, UK.
81 Hannon, J.B., Kodambaka, S., Ross, F.M., and Tromp, R.M. (2006) The influence of the surface migration of gold on the growth of silicon nanowires. *Nature*, **440**, 69.
82 Murayama, M. and Nakayama, T. (1994) Chemical trend of band offset at wurtzite/zinc-blende heterocrystalline semiconductor interfaces. *Phys. Rev. B*, **49**, 4710.
83 Yeh, C.Y., Lu, Z.W., Froyen, S., and Zunger, A. (1992) Zinc-blende-wurtzite polytypism in semiconductors. *Phys. Rev. B*, **46**, 10086.
84 Caroff, P., Dick, K.A., Johansson, J., Messing, M.E., Deppert, K., and Samuelson, L. (2008) Controlled polytypic and twin-plane superlattices in III–V nanowires. *Nat. Nanotechnol.*, **4**, 50.
85 Bolinsson, J., Ouattara, L., Hofer, W.A., Skoeld, N., Lundgren, E., Gustafsson, A., and Mikkelsen, A. (2009) Direct observation of atomic scale surface relaxation in ortho twin structures in GaAs by XSTM. *J. Phys.: Condens. Matter.*, **21**, 055404.

86 Akiyama, T., Sano, K., Nakamura, K., and Ito, T. (2006) An empirical potential approach to wurtzite-zinc-blende polytypism in group III–V semiconductor nanowires. *Jpn. J. Appl. Phys.*, **45**, L275.

87 Koguchi, M., Kakibayashi, H., Yazawa, M., Hiruma, K., and Katsuyama, T. (1992) Crystal structure change of GaAs and InAs whiskers from zinc-blende to wurtzite type. *Jpn. J. Appl. Phys.*, **31**, 2061.

88 Dick, K.A., Caroff, P., Bolinsson, J., Messing, M.E., Johansson, J., Deppert, K., Wallenberg, L.R., and Samuelson, L. (2010) Control of III–V nanowire crystal structure by growth parameter tuning. *Semicond. Sci. Technol.*, **25**, 024009.

89 Glas, F., Harmand, J.-C., and Patriarche, G. (2007) Why does wurtzite form in nanowires of III–V zinc blende semiconductors?. *Phys. Rev. Lett.*, **99**, 146101.

90 Glas, F., Patriarche, G., and Harmand, J.C. (2010) Growth, structure and phase transitions of epitaxial nanowires of III–V semiconductors. *J. Phys. Conf. Ser.*, **209**, 012002.

91 Algra, R.E., Verheijen, M.A., Borgstroem, M.T., Feiner, L.F., Immink, G., van Enckevort, W.J.P., Vlieg, E., and Bakkers, E.P.A.M. (2008) Twinning superlattices in indium phosphide nanowires. *Nature*, **456**, 369.

92 Hiruma, K., Yazawa, M., Haraguchi, K., Ogawa, K., Katsuyama, T., Koguchi, M., and Kakibayashi, H. (1993) GaAs free-standing quantum-size wires. *J. Appl. Phys.*, **74**, 3162.

93 Krishnamachari, U., Borgstrom, M., Ohlsson, B.J., Panev, N., Samuelson, L., Seifert, W., Larsson, M.W., and Wallenberg, L.R. (2004) Defect-free InP nanowries grown in [001] direction on InP(001). *Appl. Phys. Lett.*, **85**, 2077.

94 Mikkelsen, A., Skoeld, N., Ouattara, L., and Lundgren, E. (2006) Nanowire growth and dopants studied by cross-sectional scanning tunneling microscopy. *Nanotechnology*, **17**, S362.

95 Ross, F.M., Tersoff, J., and Reuter, M.C. (2005) Sawtooth faceting in silicon nanowires. *Phys. Rev. Lett.*, **95**, 146104.

96 Kamins, T.I., Li, X., Williams, R.S., and Liu, X. (2004) Growth and structure of chemically vapor deposited Ge nanowires on Si substrates. *Nano Lett.*, **4**, 503.

97 Bakkers, E.P.A.M., van Dam, J.A., De Franceschi, S., Kouwenhoven, L.P., Kaiser, M., Verheijen, M., Wondergem, H., and van der Sluis, P. (2004) Epitaxial growth of InP nanowires on germanium. *Nat. Mater.*, **3** 769.

98 Mårtensson, T., Svensson, C.P.T., Wacaser, B.A., Larsson, M.W., Seifert, W., Deppert, K., Gustafsson, A., Wallenberg, L.R., and Samuelson, L. (2004) Epitaxial III–V nanowires on silicon. *Nano Lett.*, **4**, 1987.

99 Kuykendall, T., Pauzauskie, P.J., Zhang, Y., Goldberger, J., Sirbuly, D., Denlinger, J., and Yang, P. (2004) Crystallographic alignment of high-density gallium nitride nanowire arrays. *Nat. Mater.*, **3**, 524.

100 Bjork, M.T., Thelander, C., Hansen, A.E., Jensen, L.E., Larsson, M.W., Wallenberg, L.R., and Samuelson, L. (2004) Few-electron quantum dots in nanowires. *Nano Lett.*, **4**, 1621.

101 Lauhon, L.J., Gudikson, M.S., Wang, D., and Lieber, C.M. (2002) Epitaxial core-shell and core-multishell nanowire heterostructures. *Nature*, **420**, 57.

102 Milliron, D.J., Hughes, S.M., Cui, Y., Manna, L., Li, J., Wang, L.W., and Alivisatos, A.P. (2004) Colloidal nanocrystal heterostructures with linear and branched topology. *Nature*, **430**, 190.

103 Chen, Y., Ohlberg, D.A.A., and Williams, R.S. (2002) Nanowires of four epitaxial hexagonal silicides grown on Si(001). *J. Appl. Phys.*, **91**, 3213.

104 Frangis, N., Landuyt, J.V., Kaltsas, G., Travlos, A., and Nassiopolous, A.G. (1997) Growth of erbium-silicide films on (100) silicon as characterised by electron microscopy and diffraction. *J. Cryst. Growth*, **172**, 175.

105 Lee, D. and Kim, S. (2003) Formation of hexagonal Gd disilicide nanowires on Si(100). *Appl. Phys. Lett.*, **82**, 2619.

106 Lim, D.K., Bae, S.-S., Choi, J., Lee, D., Sung, D.E., Kim, S., Kim, J.K., Yeom, H. W., and Lee, H. (2008) Unidirectional Pt silicide nanowires grown on vicinal Si(100). *J. Chem. Phys.*, **128**, 094701.

107 Lin, J.-F., Bird, J.P., He, Z., Bennett, P.A., and Smith, D.J. (2004) Signatures of quantum transport in self-assembled epitaxial nickel silicide nanowires. *Appl. Phys. Lett.*, **85**, 281.

108 Preinesberger, C., Vandré, S., Kalka, T., and Daehne-Prietsch, M. (1998) Formation of dysprosium silicide wires on Si(001). *J. Phys. D: Appl. Phys.*, **31**, L43.

109 Zhou, W., Wang, S., Ji, T., Zhu, Y., Cai, Q., and Hou, X. (2006) Growth of erbium silicide nanowires on Si(001) surface studied by scanning tunneling microscopy. *Jpn. J. Appl. Phys.*, **45**, 2059.

110 Zhu, Y., Zhou, W., Wang, S., Ji, T., Hou, X., and Cai, Q. (2006) From nanowires to nanoislands: morphological evolution of erbium silicide nanostructures formed on the vicinal Si(001) surface. *J. Appl. Phys.*, **100**, 114312.

111 Crain, J.N., McChesney, J.L., Zheng, F., Gallagher, M.C., Snijders, P.C., Bissen, M., Gundelach, C., Erwin, S.C., and Himpsel, F.J. (2004) Chains of gold atoms with tailored electronic states. *Phys. Rev. B*, **69**, 125401.

112 Shen, J., Skomski, R., Klaua, M. Jenniches, H., Manoharan, S.S., Kirschner, J. (1997) Magnetism in one dimension: Fe on Cu(111). *Phys. Rev.* B, **56**, 2340.

113 Gambardella, P., Blanc, M., Brune, H., Kuhnke, K., and Kern, K. (2000) One-dimensional metal chains on Pt vicinal surfaces. *Phys. Rev. B*, **61**, 2254.

114 Miki, K., Bowler, D.R., Owen, J.H.G., Briggs, G.A.D., and Sakamoto, K. (1999) Atomically perfect bismuth lines on Si(001). *Phys. Rev. B*, **59**, 14868.

115 Omi, H. and Ogino, T. (1997) Self-assembled Ge nanowires grown on Si(113). *Appl. Phys. Lett.*, **71**, 2163.

116 Owen, J.H.G., Miki, K., and Bowler, D.R. (2006) Self-assembled nanowires on semiconductor surfaces. *J. Mater. Sci.*, **41**, 4568.

117 Wang, S.C., Yilmaz, M.B., Knox, K.R., Zaki, N., Dadap, J.I., Valla, T., Johnson, P.D., and Osgood, Jr., R.M. (2008) Electronic structure of a Co-decorated vicinal Cu(775) surface: high-resolution photoemission spectroscopy. *Phys. Rev. B*, **77**, 115448.

118 Zaki, N., Potapenko, D., Johnson, P.D., and Osgood, Jr., R.M. (2009) Atom-wide Co wires on Cu(775) at room temperature. *Phys. Rev. B*, **80**, 155419.

119 Brocks, G., Kelly, P.J., and Car, R. (1993) Adsorption of Al on Si(100): a surface polymerization reaction. *Phys. Rev. Lett.*, **70**, 2786.

120 Evans, M.M.R. and Nogami, J. (1999) Indium and gallium on Si(001): a closer look at the parallel dimer structure. *Phys. Rev. B*, **59**, 7644.

121 Albao, M.A., Evans, J.W., and Chuang, F.-C. (2009) A kinetic monte carlo study on the role of defects and detachment in the formation and growth of In chains on Si(100). *J. Phys.: Condens. Matter*, **21**, 405002.

122 Saranin, A.A., Zotov, A.V., Kotlyar, V.G., Kuyanov, I.A., Kasyanova, T.V., Nishida, A., Kishida, M., Murata, Y., Okado, H., Katayama, M., and Oura, K. (2005) Growth of thallium overlayers on a Si(100) surface. *Phys. Rev. B*, **71**, 035312.

123 Carpinelli, J.M., Weitering, H.H., Plummer, E.W., and Stumpf, R. (1996) Direct observation of a surface charge density wave. *Nature*, **381**, 398.

124 Crain, J.N., Altmann, K.N., Bromberger, C., and Himpsel, F.J. (2002) Fermi surfaces of surface states on Si(111)-Ag,Au. *Phys. Rev. B*, **66**, 205302.

125 Friend, R.H. and Jerome, D. (1979) Periodic lattice distortions and charge density waves in one- and two-dimensional metals. *J. Phys. C: Solid State Phys.*, **12**, 1441.

126 Harrison, B.C. and Boland, J.J. (2005) Real-time STM study of inter-nanowire reactions: $GdSi_2$ nanowires on Si(100). *Surf. Sci.*, **594**, 93.

127 Nogami, J., Liu, B.Z., Katkov, M.V., Ohbuchi, C., and Birge, N.O. (2001) Self-assembled rare earth silicide nanowires on Si(001). *Phys. Rev. B*, **63**, 233305.

128 Owen, J.H.G., Miki, K., Koh, H., Yeom, H.W., and Bowler, D.R. (2002) Stress relief as the driving force for self-assembled Bi nanolines. *Phys. Rev. Lett.*, **88**, 226104.

129 Wang, J.-T., Chen, C., Wang, E., and Kawazoe, Y. (2010) Magic monoatomic linear chains for Mn nanowire self-assembly on Si(001). *Phys. Rev. Lett.*, **105**, 116102.

130 Albao, M.A., Evans, M.M.R., Nogami, J., Zorn, D., Gordon, M.S., and Evans, J.W. (2005) Monotonically decreasing size distributions for one-dimensional Ga rows on Si(100). *Phys. Rev. B*, **72**, 035426.

131 Javorsky, J., Setvin, M., Ost'adal, I., Sobotik, P., and Kotrla, M. (2009) Heterogeneous nucleation and adatom detachment at one-dimensional growth of In on Si(100)(2×1). *Phys. Rev. B*, **79**, 165424.

132 Kocan, P., Jurczyszyn, L., Sobotik, P., and Ost'adal, I. (2008) Defects in the Si(100)-(2×1) surface: anchoring sites of the surface polymerization reaction of In atoms. *Phys. Rev. B*, **77**, 113301.

133 Hirjibehedin, C.F., Lutz, C.P., and Heinrich, A.J. (2006) Spin coupling in engineered atomic structures. *Science*, **312**, 1021.

134 Khajetoorians, A.A., Wiebe, J., Chilian, B., and Wiesendanger, R. (2011) Realizing all-spin-based logic operations atom by atom. *Science*, **332**, 1062.

135 Nilius, N., Wallis, T.M., and Ho, W. (2002) Development of one dimensional band structure in artificial gold chains *Science*, **297**, 1853.

136 Attard, G. and Barnes, C. (1998) *Surfaces*, Oxford Science Publications, Oxford, UK.

137 Crain, J.N., Kirakosian, A., Altmann, K.N., Bromberger, C., Erwin, S.C., McChesney, J.L., Lin, J.-L., and Himpsel, F.J. (2003) Fractional band filling in an atomic chain structure. *Phys. Rev. Lett.*, **90**, 176805.

138 Ma, D.D.D., Lee, C.S., Au, F.C.K., Tong, S.Y., and Lee, S.T. (2003) Small-diameter silicon nanowire surfaces. *Science*, **299**, 1874.

139 Delley, B. and Steigmeier, E.F. (1995) Size dependence of band gaps in silicon nanostructures. *Appl. Phys. Lett.*, **67**, 2370.

140 Read, A.J., Needs, R.J., Nash, K.J., Canham, L.T., Calcott, P.D.J., and Qteish, A. (1992) First-principles calculations of the electronic properties of silicon quantum wires. *Phys. Rev. Lett.*, **69**, 1232.

141 Niquet, Y.M., Lherbier, A., Quang, N.H., Fernandez-Serra, M.V., Blase, X., and Delerue, C. (2006) Electronic structure of semiconductor nanowires. *Phys. Rev. B*, **73**, 165319.

142 Astromskas, G., Storm, K., Karlstroem, O., Caroff, P., Borgstroem, M., and Wernersson, L.-E. (2010) Doping incorporation in InAs nanowires characterized by capacitance measurements. *J. Appl. Phys.*, **108**, 054306.

143 Xie, P., Hu, Y., Fang, Y., Huang, J., and Lieber, C.M. (2009) Diameter-dependent dopant location in silicon and germanium nanowires. *Proc. Natl. Acad. Sci.*, **106**, 15254.

144 Yang, C., Zhong, Z., and Lieber, C.M. (2005) Encoding electronic properties by synthesis of axial modulation-doped silicon nanowires. *Science*, **310**, 1304.

145 Bjoerk, M.T., Schmid, H., Knoch, J., Riel, H., and Riess, W. (2009) Donor deactivation in silicon nanostructures. *Nat. Nanotechnol.*, **4**, 103.

146 Moench, W. (1995) *Semiconductor Surfaces and Interfaces*, 2nd edn, Springer-Verlag, Berlin, Heidelberg, New York.

147 Calarco, R., Marso, M., Richter, T., Aykanat, A.I., Meijers, R., v.d. Hart, A., Stoica, T., and Lueth, H. (2005) Size-dependent photoconductivity in MBE-grown GaN nanowires. *Nano Lett.*, **5**, 981.

148 Adachi, S. (1992) *Physical Properties of III–V Semiconductor Compounds*, Wiley-VCH, New York.

149 Diarra, M., Niquet, Y.-M., Delerue, C., and Allan, G. (2007) Ionization energy of donor and acceptor impurities in semiconductor nanowires: importance of dielectric confinement. *Phys. Rev. B*, **75**, 045301.

150 Massalski, T.B. (1990) *Binary Alloy Phase Diagrams*, 2nd edn, ASM International, Materials Park, Ohio, USA.

151 Baletto, F. and Ferrando, R. (2005) Structural properties of nanoclusters: energetic, thermodynamic, and kinetic effects. *Rev. Mod. Phys.*, **77**, 371.

152 Brack, M. (1993) The physics of simple metal clusters: self-consistent jellium model and semiclassical approaches. *Rev. Mod. Phys.*, **65**, 677.

153 Clemenger, K. (1985) Ellipsoidal shell structure in free-electron metal clusters. *Phys. Rev. B*, **32**, 1359.

154 Dong, Y. and Springborg, M. (2007) Unbiased determination of structural and electronic properties of gold clusters with up to 58 atoms. *J. Phys. Chem. C*, **111**, 12528.

155 Ferrando, R., Fortunelli, A., and Rossi, G. (2005) Quantum effects on the structure of pure and binary metallic nanoclusters. *Phys. Rev. B*, **72**, 085449.

156 Gruene, P., Rayner, D.M., Redlich, B., van der Meer, A.F.G., Lyon, J.T., Meijer, G., and Fielicke, A. (2008) Structures of neutral Au_7, Au_{19}, and Au_{20} clusters in the gas phase. *Science*, **321**, 674.

157 Han, Y. and Liu, D.-J. (2010) Shell structure and phase relations in electronic properties of metal nanowires from an electron-gas model. *Phys. Rev. B*, **82**, 125420.

158 deHeer, W.A. (1993) The physics of simple metal clusters: experimental aspects and simple models. *Rev. Mod. Phys.*, **65**, 611.

159 Aiken III, J.D. and Finke, R.G. (1999) A review of modern transition-metal nanoclusters: their synthesis, characterization, and applications in catalysis. *J. Mol. Catal. A: Chem.*, **145**, 1.

160 Itoh, M., Kumar, V., Adschiri, T., and Kawazoe, Y. (2009) Comprehensive study of sodium, copper, and silver clusters over a wide range of sizes $2 \leq N \leq 75$. *J. Chem. Phys.*, **131**, 174510.

161 Knight, W.D., Clemenger, K., de Heer, W.A., Saunders, W.A., Chou, M.Y., and Cohen, M.L. (1984) Electronic shell structure and abundances of sodium clusters. *Phys. Rev. Lett.*, **52**, 2141.

162 Martin, T.P. (1996) Shells of atoms. *Phys. Rep.*, **273**, 199.

163 Martin, T.P. (2000) From atoms to solids. *Solid State Ionics*, **131**, 3.

164 Rossi, G., Rapallo, A., Mottet, C., Fortunelli, A., Baletto, F., and Ferrando, R. (2004) Magic polyicosahedral core-shell clusters. *Phys. Rev. Lett.*, **93**, 105503.

165 Jain, P.K., Huang, X., El-Sayed, I.H., and El-Sayed, M.A. (2007) Review of some interesting surface plasmon resonance-enhanced properties of noble metal nanoparticles and their applications to biosystems. *Plasmonics*, **2**, 107.

166 Ozbay, E. (2006) Plasmonics: merging photonics and electronics at nanoscale dimensions. *Science*, **311**, 189.

167 Dubertret, B., Skourides, P., Norris, D.J., Noireaux, V., Brivanlou, A.H., and Libchaber, A. (2002) *In vivo* imaging of quantum dots encapsulated in phospholipid micelles. *Science*, **298**, 1759.

168 Shi, J., Votruba, A.R., Farokhzad, O.C., and Langer, R. (2010) Nanotechnology in drug delivery and tissue engineering: from discovery to applications. *Nano Lett.*, **10**, 3223.

169 Chang, L.Y., Barnard, A.S., Gontard, L.C., and Dunin-Borkowski, R.E. (2010) Resolving the structure of active sites on platinum catalytic nanoparticles. *Nano Lett.*, **10**, 3073.

170 Gontard, L.C., Chang, L.-Y., Hetherington, C.J.D., Kirkland, A.I., Ozkaya, D., and Dunin-Borkowski, R.E. (2007) Aberration-corrected imaging of active sites on industrial catalyst nanoparticles. *Angew. Chem. Int. Ed.*, **46**, 3683.

171 Kaden, W.E., Wu, T., Kunkel, W.A., and Anderson, S.L. (2009) Electronic structure controls reactivity of size-selected Pd-clusters adsorbed on TiO_2 surfaces. *Science*, **326**, 826.

172 Meyer, R., Lemire, C., Shaikhutdinov, S.K., and Freund, H.-J. (2004) Surface chemistry of catalysis by gold. *Gold Bull.*, **37**, 72.

173 Zope, B.N., Hibbitts, D.D., Neurock, M., and Davis, R.J. (2010) Reactivity of the gold/water interface during selective oxidation catalysis. *Science*, **330**, 74.

174 Farmer, J.A. and Campbell, C.T. (2010) Ceria maintains smaller metal catalyst particles by strong metal-support bonding. *Science*, **329**, 933.

175 Vajda, S., Pellin, M.J., Greeley, J.P., Marshall, C.L., Curtiss, L.A., Ballentine, G. A., Elam, J.W., Catillon-Mucherie, S., Redfern, P.C., Mehmood, F., and Zapol, P. (2009) Subnanometer platinum clusters as highly active and selective catalysts for the oxidative dehydrogenation of propane. *Nat. Mater.*, **8**, 213.

176 Yoon, B., Haekkinen, H., Landman, U., Woerz, A.S., Antonietti, J.-M., Abbet, S., Judai, K., and Heiz, U. (2005) Charging effects on bonding and catalyzed oxidation of CO on Au_8 clusters on MgO. *Science*, **307**, 403.

177 N'Diaye, A.T., Bleikamp, S., Feibelman, P.J., and Michely, T. (2006) Two-dimensional Ir cluster lattice on a graphene Moire on Ir(111). *Phys. Rev. Lett.*, **97**, 215501.

178 Kotlyar, V.G., Zotov, A.V., Saranin, A.A., Kasyanova, T.V., Cherevik, M.A., Pisarenko, I.V., and Lifshitz, V.G. (2002) Formation of the ordered array of Al magic clusters on Si(111)(7×7). *Phys. Rev. B*, **66**, 165401.

179 Vitali, L., Ramsey, M.G., and Netzer, F.P. (1999) Nanodot formation on the Si(111)-(7×7) surface by adatom trapping. *Phys. Rev. Lett.*, **83**, 316.

180 Bolotin, K.I., Kuemmeth, F., Pasupathy, A.N., and Ralph, D.C. (2004) Metal-nanoparticle single-electron transistors fabricated using electromigration. *Appl. Phys. Lett.*, **84**, 3154.

181 Heiz, U., Vayloyan, A., and Schumacher, E. (1997) A new cluster source for the generation of binary metal clusters. *Rev. Sci. Inst.*, **68**, 3718.

182 von Issendorff, B. and Palmer, R.E. (1999) A new high transmission infinite range mass selector for cluster and nanoparticle beams. *Rev. Sci. Inst.*, **70**, 4497.

183 Jensen, P. (1999) Growth of nanostructures by cluster deposition: experiments and simple models. *Rev. Mod. Phys.*, **71**, 1695.

184 Larsen, R.A., Neoh, S.K., and Herschbach, D.R. (1974) Seeded supersonic alkali atom beams. *Rev. Sci. Inst.*, **45**, 1511.

185 Magnusson, M.H., Deppert, K., Malm, J.-O., Bovin, J.-O., and Samuelson, L. (1999) Gold nanoparticles: production, reshaping, and thermal charging. *J. Nanopart. Res.*, **1**, 243.

186 Magnusson, M.H., Deppert, K., Malm, J.O., Bovin, J.-O., and Samuelson, L. (1999) Size-selected gold nanoparticles by aerosol technology. *Nanostruct. Mater.*, **12**, 45.

187 Scheibel, H.G. and Porstendoerfer, J. (1983) Generation of monodisperse Ag- and NaCl- aerosols with particle diameters between 2 and 300 nm. *J. Aerosol. Sci.*, **14**, 113.

188 Kappes, M.M., Kunz, R.W., and Schumacher, E. (1982) Production of large sodium clusters (Na_x, $x<65$) by seeded beam expansion. *Chem. Phys. Lett.*, **91**, 413.

189 Boyen, H.-G., Kaestle, G., Weigl, F., Ziemann, P., Schmid, G., Garnier, M.G., and Oelhafen, P. (2001) Chemically induced metal-to-insulator transition in Au_{55} clusters: effect of stabilizing ligands on the electronic properties of nanoparticles. *Phys. Rev. Lett.*, **87**, 276401.

190 Schmid, G. (1992) Large clusters and colloids. Metals in the embryonic state. *Chem. Rev.*, **92**, 1709.

191 Zhang, H., Hartmann, U., and Schmid, G. (2004) Energy-level splitting of ligand-stablized Au_{55} clusters observed by scanning tunneling spectroscopy. *Appl. Phys. Lett.*, **84**, 1543.

192 Jung, Y.S., Lee, J.H., Lee, J.Y., and Ross, C.A. (2010) Fabrication of diverse metallic nanowire arrays based on block copolymer self-assembly. *Nano Lett.*, **10**, 3722.

193 Kaestle, G., Boyen, H.-G., Weigl, F., Lengl, G., Herzog, T., Ziemann, P., Riethmueller, S., Mayer, O., Hartmann, C., Spatz, J.P., Moeller, M., Ozawa, M., Banhart, F., Garnier, M.G., and Oelhafen, P. (2003) Micellar nanoreactors – preparation and characterization of hexagonally ordered arrays of metallic nanodots. *Adv. Funct. Mater.*, **13**, 853.

194 Wu, K., Fujikawa, Y., Nagao, T., Hasegawa, Y., Nakayama, K.S., Xue, Q.K., Wang, E.G., Briere, T., Kumar, V., Kawazoe, Y., Zhang, S.B., and Sakurai, T. (2003) Na adsorption on the Si(111)-(7×7) surface: from two-dimensional gas to nanocluster array. *Phys. Rev. Lett.*, **91**, 126101.

195 Li, J.-L., Jia, J.-F., Liang, X.-J., Liu, X., Wang, J.-Z., Xue, Q.-K., Li, Z.-Q., Tse, J. S., Zhang, Z., and Zhang, S.B. (2002) Spontaneous assembly of perfectly ordered identical-size nanocluster arrays. *Phys. Rev. Lett.*, **88**, 066101.

196 Li, S.-C., Jia, J.-F., Dou, R.-F., Xue, Q.K., Batyrev, I.G., and Zhang, S.B. (2004) Borderline magic clustering: the fabrication of tetravalent Pb-cluster arrays on Si(111)-(7×7). *Phys. Rev. Lett.*, **93**, 116103.

197 Li, Z.Y., Young, N.P., DiVece, M., Palomba, S., Palmer, R.E., Bleloch, A.L., Curley, B.C., Johnston, R.L., Jiang, J., and Yuan, J. (2008) Three-dimensional atomic-scale structure of size selected gold nanoclusters. *Nature*, **451**, 46.

198 Martin, T.P. (1983) Alkali halide clusters and microcrystals. *Phys. Rep.*, **95**, 167.

199 Shibata, T., Bunker, B.A., Zhang, Z., Meisel, D., Vardeman II, C.F., and Gezelter, J.D. (2002) Size-dependent spontaneous alloying of Au-Ag nanoparticles. *J. Am. Chem. Soc.*, **124**, 11989.

200 Christensen, A., Stoltze, P., and Norskov, J.K. (1995) Size dependence of phase separation in small bimetallic clusters. *J. Phys.: Condens. Matter.*, **7**, 1047.

201 Katakuse, I., Ichihara, T., Fujita, Y., Matsuo, T., Sakurai, T., and Matsuda, H. (1985) Mass distributions of copper, silver and gold clusters and electronic shell structure. *Int. J. Mass Spectrom. Ion Processes*, **67**, 229.

202 Haberland, H., Hippler, T., Donges, J., Kostko, O., Schmidt, M., and von Issendorff, B. (2005) Melting of sodium clusters: where do the magic numbers come from?. *Phys. Rev. Lett.*, **94**, 035701.

203 Martin, T.P., Bergmann, T., Goehlich, H., and Lange, T. (1990) Observation of electronic shells and shells of atoms in large Na clusters. *Chem. Phys. Lett.*, **172**, 209.

204 Chen, Z.Y. and Castleman, A.W. (1993) Growth of titanium nitride: from clusters to microcrystals. *J. Chem. Phys.*, **98**, 231.

205 Doye, J.P.K., Wales, D.J., and Berry, R.S. (1995) The effect of the range of the potential on the structures of clusters. *J. Chem. Phys.*, **103**, 4234.

206 Baletto, F., Mottet, C., and Ferrando, R. (2002) Freezing of silver nanodroplets. *Chem. Phys. Lett.*, **354**, 82.

207 Harrison, P. (2009) *Quantum Wells, Wires and Dots*, John Wiley and Sons, Chichester, UK.

208 Bawendi, M.G., Steigerwald, M.L., and Brus, L.E. (1990) The quantum mechanics of larger semiconductor clusters ("quantum dots"). *Annu. Rev. Phys. Chem.*, **41**, 477.

209 Murray, C.B., Kagan, C.R., and Bawendi, M.G. (2000) Synthesis and characterization of monodisperse nanocrystals and close-packed nanocrystal assemblies. *Annu. Rev. Mater. Res.*, **30**, 545.

210 Redl, F.X., Cho, K.-S., Murray, C.B., and O'Brien, S. (2003) Three-dimensional binary superlattices of magnetic nanocrystals and semiconductor quantum dots. *Nature*, **423**, 968.

211 Collier, C.P., Vossmeyer, T., and Heath, J.R. (1998) Nanocrystal superlattices. *Annu. Rev. Phys. Chem.*, **49**, 371.

212 Peng, X., Manna, L., Yang, W., Wickham, J., Scher, E., Kadavanich, A., and Alivisatos, A.P. (2000) Shape control of CdSe nanocrystals. *Nature*, **404**, 59.

213 Liu, Q.K.K., Moll, N., Scheffler, M., and Pehlke, E. (1999) Equilibrium shapes and energies of coherent strained InP islands. *Phys. Rev. B*, **60**, 17008.

214 Voigtlaender, B. (2001) Fundamental processes in Si/Si and Ge/Si epitaxy studied by scanning tunneling microscopy during growth. *Surf. Sci. Rep.*, **43**, 127.

215 Li, J. and Wang, L.-W. (2003) Shape effects on electronic states of nanocrystals. *Nano Lett.*, **3**, 1357.

216 Banin, U., Cao, Y.W., Katz, D., and Millo, O. (1999) Identification of atomic-like electronic states in indium arsenide nanocrystal quantum dots. *Nature*, **400**, 542.

217 Grundmann, M., Christen, J., Ledentsov, N.N., Boehrer, J., Bimberg, D., Ruvimov, S.S., Werner, P., Richter, U., Goesele, U., Heydenreich, J., Ustinov, V.M., Egorov, A.Y., Zhukov, A.E., Kop'ev, P.S., and Alferov, Z.O. (1995) Ultranarrow luminescence lines from single quantum dots. *Phys. Rev. Lett.*, **74**, 4043.

218 Kamat, P.V. (2008) Quantum dot solar cells. semiconductor nanocrystals as light harvesters. *J. Phys. Chem. C*, **112**, 18737.

219 Schaller, R.D., Agranovich, V.M., and Klimov, V.I. (2005) High-efficiency carrier multiplication through direct photogeneration of multi-excitons via virtual single-exciton states. *Nat. Phys.*, **1**, 189.

220 Tatebayashi, J., Nishioka, M., and Arakawa, Y. (2001) Over 1.5 micrometer light emission from InAs quantum dots embedded in InGaAs strain-reducing layer grown by metalorganic chemical vapor deposition. *Appl. Phys. Lett.*, **78**, 3469.

221 LaMer, V. and Dinegar, R.H. (1950) Theory, production and mechanism of formation of monodispersed hydrosols. *J. Am. Chem. Soc.*, **72**, 4847.

222 Murray, C.B., Norris, D.J., and Bawendi, M.G. (1993) Synthesis and characterization of nearly monodisperse CdE (E=S,Se,Te) semiconductor nanocrystallites. *J. Am. Chem. Soc.*, **115**, 8706.

223 Brus, L.E. (1984) Electron–electron and electron–hole interactions in small semiconductor crystallites: the size dependence of the lowest excited electronic state. *J. Chem. Phys.*, **80**, 4403.

224 Wang, L.-W. and Zunger, A. (1996) Pseudopotential calculations of nanoscale CdSe quantum dots. *Phys. Rev. B*, **53**, 9579.

225 Bruchez, M., Moronne, M., Gin, P., Weiss, S., and Alivisatos, A.P. (1998) Semiconductor nanocrystals as fluorescent biological labels. *Science*, **281**, 2013.

226 Trindade, T., O'Brien, P., and Pickett, N.L. (2001) Nanocrystalline semiconductors; synthesis, properties and perspectives. *Chem. Mater.*, **13**, 3843.

227 Hines, M.A. and Guyot-Sionnest, P. (1998) Bright UV-blue luminescent colloidal ZnSe nanocrystals. *J. Phys. Chem. B*, **102**, 3655.

228 Peng, X. (2003) Mechanisms for the shape-control and shape-evolution of colloidal semiconductor nanocrystals, *Adv. Mater.*, **15**, 459.

229 Peng, Z.A. and Peng, X. (2001) Formation of high-quality CdTe, CdSe, and CdS nanocrystals using CdO as precursor, *J. Am. Chem. Soc.*, **123**, 183.

230 Yin, Y. and Alivisatos, A.P. (2005) Colloidal nanocrystal synthesis and the organic–inorganic interface, *Nature*, **437**, 664.

231 Kim, S., Fisher, B., Eisler, H.-J., and Bawendi, M. (2003) Type-II quantum dots: CdTe/CdSe (core/shell) and CdSe/ZnTe(core/shell) heterostructures, *J. Amer. Chem. Soc.*, **125**, 11466.

232 Zhang, J., Tang, Y., Lee, K., and Ouyang, M. (2010) Nonepitaxial growth of hybrid core-shell nanostructures with large lattice mismatches, *Science*, **327**, 1634.

233 Reiss, H. (1951) The growth of uniform colloidal dispersions, *J. Chem. Phys.*, **19**, 482.

234 Peng, X., Wickham, J., and Alivisatos, A.P. (1998) Kinetics of II-VI and III-V colloidal semiconductor nanocrystal growth: "focusing" of size distributions, *J. Am. Chem. Soc.*, **120**, 5343.

235 Burda, C., Chen, X., Narayanan, R., and El-Sayed, M.A. (2005) Chemistry and properties of nanocrystals of different shapes, *Chem. Rev.*, **105**, 1025.

236 Chen, S., Wang, Z.L., Ballato, J., Foulger, S.H., and Carroll, D.L. (2003) Monopod, bipod, tripod, and tetrapod gold nanocrystals, *J. Am. Chem. Soc.*, **125**, 16186.

237 Lee, K., Kim, M., and Kim, H. (2010) Catalytic nanoparticles being facet-controlled, *J. Mater. Chem.*, **20**, 3791.

238 Manna, L., Milliron, D.J., Meisel, A., Scher, E.C., and Alivisatos, A.P. (2003) Controlled growth of tetrapod-branched inorganic nanocrystals, *Nat. Mat.*, **2**, 382.

239 Peng, Z.A. and Peng, X. (2002) Nearly monodisperse and shape-controlled CdSe nanocrystals via alternative routes: nucleation and growth, *J. Am. Chem. Soc.*, **124**, 3343.

240 Sau, T.K. and Murphy, C.J. (2004) Room temperature, high-yield synthesis of multiple shapes of gold nanoparticles in aqueous solution, *J. Am. Chem. Soc.*, **126**, 8648.

241 Scher, E.C., Manna, L., and Alivisatos, A.P. (2003) Shape control and applications of nanocrystals, *Phil. Trans. Royal Soc. A*, **361**, 241.

242 Mullins, J.W. (2001) *Crystallization*, 2nd edn, Butterworth-Heinemann, Oxford, UK.

243 Manna, L., Scher, E.C., and Alivisatos, A.P. (2000) Synthesis of soluble and processable rod-, arrow-, teardrop-, and tetrapod shaped CdSe nanocrystals, *J. Am. Chem. Soc.*, **122**, 12700.

244 Puntes, V.F., Krishnan, K.M., and Alivisatos, A.P. (2011) Colloidal nanocrystal shape and size control: the case of cobalt, *Science*, **291**, 2115.

245 Ma, Y., Li, W., Cho, A.C., Li, Z., Yu, T., Zeng, J., Xie, Z., and Xia, Y. (2010) Au@Ag core-shell nanocubes with finely tuned and well-controlled sizes, shell thicknesses and optical properties, *ACS Nano*, **4**, 6725.

246 Portavoce, A., Hull, R., Reuter, M.C., and Ross, F.M. (2007) Nanometer-scale control of single quantum dot nucleation through focussed ion-beam implantation, *Phys. Rev. B*, **76**, 235301.

247 Solomon, G.S., Trezza, J.A., Marshall, A.F., and Harris, J.S. (1996) Vertically aligned and electronically coupled growth induced InAs islands in GaAs, *Phys. Rev. Lett.*, **76**, 952.

248 Bortoleto, J.R.R., Gazoto, A., Brasil, M.J.S.P., Meneses, E.A., and Cotta, M.A. (2010) Nucleation and growth evolution of InP dots on InGaP/GaAs, *J. Phys.: D. Appl. Phys.*, **43**, 285301.

249 Daudin, B., Widmann, F., Feuillet, G., Samson, Y., Arlery, M., and Rouvière, J.L. (1997) Strainski–Krastanov growth mode during the molecular beam epitaxy of highly strained GaN, *Phys. Rev. B*, **56**, R7069.

250 Eaglesham, D.J. and Cerullo, M. (1990) Dislocation-free Stranski–Krastanow growth of Ge on Si(100), *Phys. Rev. Lett.*, **64**, 1943.

251 Grundmann, M., Stier, O., and Bimberg, D. (1995) InAs/GaAs pyramidal quantum dots: strain distribution, optical phonons, and electronic structure, *Phys. Rev. B*, **52**, 11969.

252 Moison, J.M., Houzay, F., Barthe, F., Leprince, L., André, E., and Vatel, O. (1994) Self-organized growth of regular nanometer-scale InAs dots on GaAs, *Appl. Phys. Lett.*, **64**, 196.

253 Wang, L.G., Kratzer, P., Moll, N., and Scheffler, M. (2000) Size, shape, and stability of InAs quantum dots on the GaAs(001) substrate, *Phys. Rev. B*, **62**, 1897.

254 Georgsson, K., Carlsson, N., Samuelson, L., Seifert, W., and Wallenberg, L.R. (1995) Transmission electron microscopy investigation of the morphology of InP Stranski–Krastanov islands grown by metalorganic chemical vapor deposition, *Appl. Phys. Lett.*, **67**, 2981.

255 Hatami, F., Ledentsov, N.N., Grundmann, M., Boehrer, J., Heinrichsdorff, F., Beer, M., Bimberg, D., Ruvimiv, S.S., Werner, P., Goesele, U., Heydenreich, J.,

Richter, U., Ivanov, S.V., Meltser, B.Y., Kop'ev, P.S., and Alferov, Z.I. (1995) Radiative recombination in type-II GaSb/GaAs quantum dots, *Appl. Phys. Lett.*, **67**, 656.

256 Ponchet, A., LeCoree, A., L'Haridon, H., Lambert, B., and Salauen, S. (1995) Relationship between self-organization and size of InAs islands on InP(001) grown by gas source molecular beam epitaxy, *Appl. Phys. Lett.*, **67**, 1850.

257 Floro, J.A., Chason, E., Sinclair, M.B., Freund, L.B., and Lucadamo, G.A. (1998) Dynamic self-organization of strained islands during SiGe epitaxial growth, *Appl. Phys. Lett.*, **73**, 951.

258 Gu, Y., Yang, T., Ji, H., Xu, P., and Wang, Z. (2011) Redshift and discrete energy level separation of self-assembled quantum dots induced by strain-reducing layer, *J. Appl. Phys.*, **109**, 064320.

259 Ustinov, V.M., Maleev, N.A., Zhukov, A.E., Kovsh, A.R., Egorov, A.Y., Lunev, A. V., Volovik, B.V., Krestnikov, I.L., Musikhin, Y.G., Bert, N.A., Kop'ev, P.S., Alferov, Z.I., Ledentsov, N.N., and Bimberg, D. (1999) InAs/InGaAs quantum dot structure on GaAs substrates emitting at 1.3 μm, *Appl. Phys. Lett.*, **74**, 2815.

260 Meyer, J.A., Schmid, P., and Behm, R.J. (1995) Effect of layer-dependent adatom mobilities in heteroepitaxial metal film growth: Ni/Ru(0001), *Phys. Rev. Lett.*, **74**, 3864.

261 Pehlke, E., Moll, N., Kley, A., and Scheffler, M. (1997) Shape and stability of quantum dots, *Appl. Phys. A*, **65**, 525.

262 Ratsch, C. and Zangwill, A. (1993) Equilibrium theory of the Stranski–Krastanov epitaxial morphology, *Surf. Sci.*, **293**, 123.

263 Ibach, H. (2006) *Physics of Surfaces and Interfaces*, Springer-Verlag, Berlin, New York.

264 Kukta, R.V. and Freund, L.B. (1997) Minimum energy configuration of epitaxial material clusters on a lattice-mismatched substrate, *J. Mech. Phys. Solids*, **45**, 1835.

265 Shchukin, V.A. and Bimberg, D. (1999) Spontaneous ordering of nanostructures on crystal surfaces, *Rev. Mod. Phys.*, **71**, 1125.

266 Daruka, I. and Barabasi, A.-L. (1997) Dislocation-free island formation in heteroepitaxial growth: a study at equilibrium, *Phys. Rev. Lett.*, **79**, 3708.

267 Daruka, I., Tersoff, J., and Barabasi, A.-L. (1999) Shape transition in growth of strained islands, *Phys. Rev. Lett.*, **82**, 2753.

268 Kaestner, M. and Voigtlaender, B. (1999) Kinetically self-limiting growth of Ge-islands on Si(001), *Phys. Rev. Lett.*, **82**, 2745.

269 Spencer, B.J., Voorhees, P.W., and Davis, S.H. (1991) Morphological instability in epitaxially strained dislocation-free solid films, *Phys. Rev. Lett.*, **67**, 3696.

270 Spencer, B.J., Voorhees, P.W., and Davis, S.H. (1993) Morphological instability in epitaxially strained dislocation-free solid films: linear stability theory, *J. Appl. Phys.*, **73**, 4955.

271 Gray, J.L., Hull, R., Lam, C.-H., Sutter, P., Means, J., and Floro, J. A. (2005) Beyond the heteroepitaxial quantum dot: self-assembling complex nanostructures controlled by strain and growth kinetics, *Phys. Rev. B*, **72**, 155323.

272 Gray, J.L., Singh, N., Elzey, D.M., Hull, R., and Floro, J.A. (2004) Kinetic size selection mechanism in heteroepitaxial quantum dot molecules, *Phys. Rev. Lett.*, **92**, 1335504.

273 Medeiros-Ribeiro, G., Bratkovski, A.M., Kamins, T.I., Ohlberg, D.A.A., and Williams, R.S. (1998) Shape transition of germanium nanocrystals on a silicon (001) surface from pyramids to domes, *Science*, **279**, 353.

274 Andreev, A.D., Downes, J.R., Faux, D.A., and O'Reilly, E.P. (1999) Strain distributions in quantum dots of arbitrary shape. *J. Appl. Phys.*, **86**, 297.

275 Mo, Y.-W., Savage, D.E., Swartzentruber, B.S., and Lagally, M.G. (1990) Kinetic pathway in Stranski–Krastanov growth of Ge on Si(001), *Phys. Rev. Lett.*, **65**, 1020.

276 Ross, F.M., Tersoff, J., and Tromp, R.M. (1998) Coarsening of self-assembled Ge quantum dots on Si(001), *Phys. Rev. Lett.*, **80**, 984.

277 Bartlett, P., Ottewill, R.H., and Pusey, P.N. (1992) Superlattice formation in binary mixtures of hard-sphere colloids, *Phys. Rev. Lett.*, **68**, 3801.

278 Eldridge, M.D., Madden, P.A., and Frenkel, D. (1993) Entropy-driven formation of a superlattice in a hard-sphere binary mixture, *Nature*, **365**, 35.

279 Evers, W.H., de Nijs, B., Filion, L., Castillo, S., Dijkstra, M., and Vanmaekelbergh, D. (2010) Entropy-driven formation of binary semiconductor-nanocrystal superlattices, *Nano Lett.*, **10**, 4235.

280 Hunt, N., Jardine, R., and Bartlett, P. (2000) Superlattice formation in mixtures of hard-sphere colloids, *Phys. Rev. E*, **62**, 900.

281 Hynninen, A.-P., Thijssen, J.H.J., Vermolen, E.C.M., Dijkstra, M., and van Blaaderen, A. (2007) Self-assembled route for photonic crystals with a bandgap in the visible region. *Nat. Mat.*, **6**, 202.

282 Shevchenko, E.V., Talapin, D.V., Kotov, N.A., O'Brien, S., and Murray, C.B. (2006) Structural diversity in binary nanoparticle superlattices. *Nature*, **439**, 55.

283 Tersoff, J., Teichert, C., and Lagally, M.G. (1996) Self-organization in growth of quantum dot superlattices. *Phys. Rev. Lett.*, **76**, 1675.

284 Podsiadlo, P., Krylova, G.V., Demortière, A., and Shevchenko E.V. (2011) Multicomponent periodic nanoparticle superlattices. *J. Nanopart. Res.*, **13**, 15.

285 Jiang, P., Ostojic, G.N., Narat, R., Mittleman, D.M., and Colvin, V.L. (2001) The fabrication and bandgap engineering of photonic multilayers. *Adv. Mater.*, **13**, 389.

286 Hachisu, S. and Yoshimura, S. (1980) Optical demonstration of crystalline superstructures in binary mixture of latex globules. *Nature*, **283**, 188.

287 Frenkel, D. (1999) Entropy-driven phase transitions. *Phys. A*, **263**, 26.

288 Murray, M.J. and Sanders, J.V. (1980) Close-packed structures of spheres of two different sizes II. The packing density of likely arrangements. *Phil. Mag. A*, **42**, 721.

289 Saunders, A.E. and Korgel, B.A. (2005) Observation of an AB phase in bidisperse nanocrystal superlattices. *Chem. Phys. Chem.*, **6**, 61.

290 Trizac, E., Eldridge, M.D., and Madden, P.A. (1997) Stability of the AB crystal for asymmetric binary hard sphere mixtures. *Mol. Phys.*, **90**, 675.

291 Bartlett, P. and Campbell, A.I. (2005) Three-dimensional binary superlattices of oppositely charged colloids. *Phys. Rev. Lett.*, **95**, 128302.

292 Sanders, J.V. and Murray, M.J. (1978) Ordered arrangements of spheres of two different sizes in opal. *Nature*, **275**, 201.

293 Shevchenko, E.V., Ringler, M., Schwemer, A., Talapin, D.V., Klar, T.A.., Rogach, A.L., Feldmann, J. and Alivisatos, A.P. (2008) Self-assembled binary superlattices of CdSe and Au nanocrystals and their fluorescence properties. *J. Amer. Chem. Soc.*, **130**, 3274.

294 Talapin, D.V., Shevchenko, E.V., Bodnarchuk, M.I., Ye, X., Chen, J., and Murray, C.B. (2009) Quasicrystalline order in self-assembled binary nanoparticle superlattices. *Nature*, **461**, 964.

295 Cottin, X. and Monson, P.A. (1995) Substitutionally ordered solid solutions of hard spheres. *J. Chem. Phys.*, **102**, 3354.

296 Fasolo, M. and Sollich, P. (2003) Equilibrium phase behavior of polydisperse hard spheres. *Phys. Rev. Lett.*, **91**, 068301.

297 Coleman, J.J., Young, J.D., and Garg, A. (2011) Semiconductor quantum dot lasers: a tutorial. *J. Lightwave Tech.*, **29**, 499.

298 Arakawa, Y. and Sakaki, H. (1982) Multidimensional quantum well laser and temperature dependence of its threshold current. *Appl. Phys. Lett.*, **40**, 939.

299 Gutierez, M., Herrera, M., Gonzales, D., Garcia, R., and Hopkinson, M. (2006) Role of elastic anisotropy in the vertical alignment of In(Ga)As quantum dot superlatices. *Appl. Phys. Lett.*, **88**, 193118.

300 Solomon, G.S., Trezza, J.A., Marshall, A.F., and Harris, Jr., J.S. (1996) Vertically aligned an electronically coupled growth induced InAs islands in GaAs. *Phys. Rev. Lett.*, **76**, 952.

301 Xie, Q., Madhukar, A., Chen, P., and Koboyashi, N.P. (1995) Vertically self-organized InAs quantum box islands on GaAs(100). *Phys. Rev. Lett.*, **75**, 2542.

302 Xie, Q., Chen, P., and Madhukar, A. (1994) InAs island-induced-strain driven adatom migration during GaAs overlayer growth. *Appl. Phys. Lett.*, **65**, 2051.

303 Bundy, F.P. (1980) The P,T phase and reaction diagram for elemental carbon, 1979. *J. Geophys. Res.*, **85**, 6930.

304 Bundy, F.P. and Kasper, J.S. (1967) Hexagonal diamond – a new form of carbon. *J. Chem. Phys.*, **46**, 3437.

305 Salehpour, M.R. and Satpathy, S. (1990) Comparison of electron bands of hexagonal and cubic diamond. *Phys. Rev. B*, **41**, 3048.

306 Fahy, S., Louie, S.G., and Cohen, M.L. (1986) Theoretical total-energy study of the transformation of graphite into hexagonal diamond. *Phys. Rev. B*, **35**, 7623.

307 Fahy, S., Louie, S.G., and Cohen, M.L. (1986) Pseudopotential total-energy study of the transition from rhombohedral graphite to diamond. *Phys. Rev. B*, **34**, 1191.

308 Robertson, J. and O'Reilly, E.P. (1986) Electronic and atomic structure of amorphous carbon. *Phys. Rev. B*, **35**, 2946.

309 Hofsaess, H., Feldermann, H., Merk, R., Sebastian, M., and Ronning, C. (1998) Cylindrical spike model for the formation of diamondlike thin films by ion deposition. *Appl. Phys. A*, **66**, 153.

310 Reinke, P. and Oelhafen, P. (1996) Thermally induced structural changes in amorphous carbon films observed with utlraviolet photoelectron spectroscopy. *J. Appl. Phys.*, **81**, 2396.

311 Graupner, R., Maier, F., Ristein, J., and Ley, L. (1998) High-resolution surface-sensitive C1s core-level spectra of clean and hydrogen-terminated diamond (100) and (111) surfaces. *Phys. Rev. B*, **57**, 12397.

312 Kroto, H.W., Heath, J.R., O'Brien, S.C., Curl, R.F., and Smalley, R.E. (1985) C_{60}: Buckminsterfullerene. *Nature*, **318**, 162.

313 Dresselhaus, M.S., Dresselhaus, G., and Fischer, J.E. (1977) Graphite intercalation compounds: electronic properties in the dilute limit. *Phys. Rev. B*, **15**, 3180.

314 Dresselhaus, M.S., Dresselhaus, G., and Eklund, P.C. (1996) *Science of Fullerenes and Carbon Nanotubes*, Academic Press, San Diego, CA, USA.

315 Taylor, R. (1999) *Lecture notes on Fullerene Chemistry: A Handbook for Chemists*, World Scientific Publishing Company, London, UK.

316 Duchamp, J.C., Demortier, A., Fletcher, K.R., Dorn, D., Iezzi, E.B., Glass, T., and Dorn, H.C. (2003) An isomer of the endohedral metallofullerene $Sc_3N@C_{80}$ with D_{5h} symmetry. *Chem. Phys. Lett.*, **375**, 655.

317 Wang, C.R., Kai, T., Tomiyama, T., Yoshida, T., Kobayashi, Y., Nishibori, E., Takata, M., Sakata, M., and Shinohara, H. (2000) C66 fullerene encaging a scandium dimer. *Nature*, **408**, 426.

318 Stevenson, S., Fowler, P.W., Heine, T., Duchamp, J.C., Rice, G., Glass, T., Harich, K., Hajdu, E., Bible, R., and Dorn, H.C. (2000) A stable non-classical metallofullerene family. *Nature*, **408**, 427.

319 Spanggaard, H. and Krebs, F.C. (2004) A brief history of the development of organic and polymeric photovoltaics. *Solar Energy Mat. Solar Cells*, **83**, 125.

320 Partha, R. and Conyers, J.L. (2009) Biomedical applications of functionalized fullerene-based nanomaterials. *Int. J. Nanomed.*, **4**, 261.

321 Oberdoerster, E. (2004) Manufactured nanomaterials (fullerenes, C_{60}) induce oxidative stress in the brain of juvenile largemouth brass. *Environ. Health*, **112**, 1058.

322 Isaacson, C.W., Kleber, M., and Field, J.A. (2009) Quantitative analysis of fullerene nanomaterials in environmental systems: a critical review. *Environ. Sci. Technol.*, **43**, 6463.

323 Kelty, C.M. (2009) Beyond implications and applications: the story of "safety by design". *Nanoethics*, **3**, 79.

324 Savolainen, K., Pylkkaenen, L., Norppa, H., Falck, G., Lindberg, H., Tuomi, T., Vippola, M., Alenius, H., Haemeri, K., Koivisto, J., Brouwer, D., Mark, D., Bard, D., Berges, M., Jankowska, E., Posniak, M., Farmer, P., Singh, R., Krombach, F., Bihari, P., Kasper, G., and Seipenbusch, M. (2010) Nanotechnologies, engineered nanomaterials and occupational health and safety – a review. *Safety Sci.*, **48**, 957.

325 Hass, J., Varchon, F., Millan-Otoya, J.E., Sprinkle, M., Sharma, N., de Heer, W.A., Berger, C., First, P.N., Magaud, L., and Conrad, E.H. (2008) Why multilayer graphene on 4H-SiC(000-1) behaves like a single sheet of graphene. *Phys. Rev. Lett.*, **100**, 125504.

326 Pacile, D., Meyer, J.C., Girit, C.O., and Zettl, A. (2008) The two-dimensional phase of boron nitride: few-atomic-layer sheets and suspended membranes. *Appl. Phys. Lett.*, **92**, 133107.

327 Golberg, D., Bando, Y., Tang, C.C., and Zhi, C.Y. (2007) Boron nitride nanotubes. *Adv. Mater.*, **19**, 2413.

328 Ci, L., Song, L., Jin, C., Jariwala, D., Wu, D., Li, Y., Srivastava, A., Wang, Z.F., Storr, K., Balicas, L., Liu, F., and Ajayan, P.M. (2010) Atomic layers of hybridized boron nitride and graphene domains. *Nat. Mater.*, **9**, 430.

329 Mak, K.F., Lee, C., Hone, J., Shan, J., and Heinz, T.F. (2010) Atomically thin MoS_2: a new direct-gap semiconductor. *Phys. Rev. Lett.*, **105**, 136805.

330 Radisavljevic, B., Radenovic, A., Brivio, J., Giacometti, V., and Kis, A. (2011) Single-layer MoS_2 transistors. *Nat. Nanotechnol.*, **6**, 147.

331 Wallace, P.R. (1947) The band theory of graphite. *Phys. Rev.*, **71**, 622.

332 Painter, G.S. and Ellis, D.E. (1970) Electronic band structure and optical properties of graphite form a variational approach. *Phys. Rev. B*, **1**, 4747.

333 Zunger, A. (1978) Self-consisted LCAO calculation of the electronic properties of graphite. I. The regular graphite lattice. *Phys. Rev. B*, **17**, 626.

334 Zunger, A. (1974) A molecular calculation of electronic properties of layered crystals. II. Periodic small cluster calculation for graphite and boron nitride. *J. Phys. C: Solid State Phys.*, **7**, 96.

335 Saito, R., Dresselhaus, G., and Dresselhaus, M.S. (1999) *Physical Properties of Carbon Nanotubes*, Imperial College Press, London, UK.

336 Odom, T.W., Huang, J.L., Kim, P., Lieber, C.M. (2000) Structure and electronic properties of carbon nanotubes. *J. Phys. Chem.* B **104**, 2794.

337 Slonczewski, J.C. and Weiss, P.R. (1958) Band structure of graphite. *Phys. Rev.*, **109**, 272.

338 Partoens, B. and Peeters, F.M. (2006) From graphene to graphite: electronic structure around the K-point. *Phys. Rev. B*, **74**, 075404.

339 McClure, J.W. (1957) Band structure of graphite and de Haas-van Alphen effect. *Phys. Rev.*, **108**, 612.

340 Hamada, N., Sawada, S., and Oshiyama, A. (1992) New one-dimensional conductors: graphitic microtubules. *Phys. Rev. Lett.*, **68**, 1579.

341 White, C.T. and Mintmire, J.W. (1998) Density of states reflects diameter in nanotube. *Nature*, **394**, 29.

342 Mintmire, J.W., Robertson, D.H., and White, C.T. (1993) Properties of fullerene nanotubules. *J. Phys. Chem. Solids*, **54**, 1835.

343 White, C.T., Robertson, D.H., and Mintmire, J.W. (1993) Helical and rotational symmetries of nanoscale graphitic tubules. *Phys. Rev. B*, **47**, 5485.

344 Iijima, S. and Ichihashi, T. (1993) Single -shell carbon nanotubes of 1-nm diameter. *Nature*, **363**, 603.

345 Yao, Z., Postma, H.W.C., Balents, L., and Dekker, C. (1999) Carbon nanotube intramolecular junctions. *Nature*, **402**, 273.

346 Andergassen, S., Meden, B.V., Schoeller, H., Splettstoesser, J., and Wegewijs, M.R. (2010) Charge transport through single molecules, quantum dots and quantum wires. *Nanotechnology*, **21**, 272001.

347 Ilani, S. and McEuen, P.L. (2010) Electron transport in carbon nanotubes. *Ann. Rev. Condens. Mat. Phys.*, **1**, 1.

348 Li, H.J., Lu, W.G., Li, J.J., Bai, X.D., and Gu, C.Z. (2005) Multichannel ballistic transport in multiwall carbon nanotubes. *Phys. Rev. Lett.*, **95**, 086601.

349 Martel, R., Schmidt, T., Shea, H.R., Hertel, T., and Avouris, P. (1998) Single- and multi-wall carbon nanotube field-effect transistors. *Appl. Phys. Lett.*, **73**, 2447.

350 Sun, D., Timmermans, M.Y., Tian,Y., Nasibulin, A.G., Kauppinen, E.I., Kishimoto, S., Mizutani, T., and Ohno, Y. (2011) Flexible high-performance carbon nanotube integrated circuits. *Nat. Nanotechnol.*, **6**, 156.

351 Andrews, R., Jacques, D., Qian, D., and Rantell, T. (2002) Multiwall carbon nanotubes: synthesis and applications. *Acc. Chem. Res.*, **35**, 1008.

352 Terrones, M. (2003) Science and technology of the twenty-first century: synthesis, properties, and applications of carbon nanotubes. *Ann. Rev. Mat. Res.*, **33**, 419.

353 Tunney, M.A. and Cooper, N.R. (2006) Effects of disorder and momentum relaxation on the intertube transport of incommensurate carbon nanotube ropes and multiwall nanotubes. *Phys. Rev. B*, **74**, 075406.

354 Lehman, J., Sanders, A., Hanssen, L., Wilthan, B., Zeng, J., and Jensen, C. (2010) Very black infrared detector from vertically aligned carbon nanotubes and electric-field poling lithium tantalate. *Nano Lett.*, **10**, 3261.

355 Bower, C., Zhu, W., Jin, S., and Zhou, O. (2000) Plasma-induced alignment of carbon nanotubes. *Appl. Phys. Lett.*, **77**, 830.

356 Chhowalla, M., Teo, K.B.K., Ducati, C., Rupesinghe, N.L., Amaratunga, G.A.J., Ferrari, A.C., Roy, D., Robertson, J., and Milne, W.I. (2001) Growth process conditions of vertically aligned carbon nanotubes using plasma enhanced chemical vapor deposition. *J. Appl. Phys.*, **90**, 5308.

357 Kocabas, C., Shim, M., and Rogers, J.A. (2006) Spatially selective guided growth of high-coverage arrays and random networks of single-walled carbon nanotubes and their integration into electronic devices. *J. Am. Chem. Soc.*, **128**, 4540.

358 Ren, Z.F., Huang, Z.P., Wang, D.Z., Wen, J.G., Xu, J.W., Wang, J.H., Calvet, L. E., Chen, J., Klemic, J.F., and Reed, M.A. (1999) Growth of a single freestanding multiwall carbon nanotube on each nanonickel dot. *Appl. Phys. Lett.*, **75**, 1086.

359 Ren, Z.F., Huang, Z.P., Xu, J.W., Wang, J.H., Bush, P., Siegal, M.P., and Provencio, P.N. (1998) Synthesis of large arrays of well-aligned carbon nanotubes on glass. *Science*, **282**, 1105.

360 Kong, J., Cassell, A.M., and Dai, H. (1998) Chemical vapor deposition of methane for single-walled carbon nanotubes. *Chem. Phys. Lett.*, **292**, 567.

361 Meyyappan, M., Delzeit, L., Cassell, A., and Hash, D. (2003) Carbon nanotube growth by PECVD: a review. *Plasma Sources Sci. Technol.*, **12**, 205.

362 Thostenson, E.T., Ren, Z., and Chou, T.-W. (2001) Advances in the science and technology of carbon nanotubes and their composites: a review. *Comp. Sci. Technol.*, **61**, 1899.

363 Nikolaev, P., Bronikowski, M.J., Bradley, R.K., Rohmund, F., Colbert, D.T., Smith, K.A., and Smalley, R.W. (1999) Gas-phase catalytic growth of single-walled carbon nanotubes from carbon monoxide. *Chem. Phys. Lett.*, **313**, 91.

364 Harutyunyan, A.R., Chen, G., Paronyan, T.M., Pigos, E.M., Kuznetsov, O.A., Hewaparakrama, K., Kim, S.M., Zakharov, D., Stach, E.A., and Sumanasekera, G.U. (2009) Preferential growth of single-walled carbon nanotubes with metallic conductivity. *Science*, **326**, 116.

365 Hersam, M.C. (2008) Progress towards monodisperse single-walled carbon nanotubes. *Nat. Nanotechnol.*, **3**, 387.

366 Zhou, S.Y., Gweon, G.-H., Graf, J., Fedorov, A.V., Spataru, C.D., Diehl, R.D., Kopelevich, Y., Lee, D.-H., Louie, S.G., and Lanzara, A. (2006) First direct observation of Dirac fermions in graphite. *Nat. Phys.*, **2**, 595.

367 Westervelt, R.M. (2008) Graphene nanoelectronics. *Science*, **320**, 324.

368 Novoselov, K.S., Jiang, D., Schedin, F., Booth, T.J., Khotkevich, V.V., Morozov, S.V., and Geim, A.K. (2005) Two-dimensional atomic crystals. *Proc. Natl. Acad. Sci.*, **102**, 10451.

369 Novoselov, K.S., Geim, A.K., Morozov, S.V., Jiang, D., Katsnelson, M.I., Grigorieva, I.V., Dubonos, S.V., and Firsov, A.A. (2005) Two-dimensional gas of massless Dirac fermions in graphene. *Nature*, **438**, 197.

370 Castro Neto, A.H., Guinea, F., Peres, N.M.R., Novoselov, K.S., and Geim, A.K. (2009) The electronic properties of graphene. *Rev. Mod. Phys.*, **81**, 109.

371 Kim, E. and Castro Neto, A.H. (2008) Graphene as an electronic membrane. *Europhys. Lett.*, **84**, 57007.

372 de Heer, W.A., Berger, C., Wu, X., Sprinkle, M., Hu, Y., Ruan, M., Stroscio, J.A., First, P.N., Haddon, R., Piot, B., Faugeras, C., Potemski, M., and Moon, J.-S. (2010) Epitaxial graphene electronic structure and transport. *J. Phys. D: Appl. Phys.*, **43**, 374007.

373 Geim, A.K. and Novoselov, K.S. (2007) The rise of graphene. *Nat. Mater.*, **6**, 183.

374 Semenoff, G.W. (1984) Condensed matter simulation of a three-dimensional anomaly. *Phys. Rev. Lett.*, **53**, 2449.

375 Zhang, Y., Tan, Y.-W., Stormer, H. L., and Kim, P. (2005) Experimental observation of the quantum Hall effect and Berry's phase in graphene. *Nature*, **438**, 201.

376 Du, X., Skachko, I., Barker, A., and Andrei, E.Y. (2008) Approaching ballistic transport on suspended graphene. *Nat. Nanotechnol.*, **3**, 491.

377 Bostwick, A., Ohta, T., Seyller, T., Horn, K., and Rotenberg, E. (2006) Quasiparticle dynamics in graphene. *Nat. Phys.*, **3**, 36.

378 Bolotin, K.I., Sikes, K.J., Jiang, Z., Klima, M., Fudenberg, G., Hone, J., Kim, P., and Stormer, H.L. (2008) Ultrahigh mobility in suspended graphene. *Solid State Commun.*, **146**, 351.

379 Miller, D.L., Kubista, K.D., Rutter, G.M., Ruan, M., de Heer, W.A., First, P.N., and Stroscio, J.A. (2009) Observing the quantization of zero mass carriers in graphene. *Science*, **324**, 924.

380 McCann, E. (2006) Asymmetry gap in the electronic band structure of bilayer graphene. *Phys. Rev. B*, **74**, 161403.

381 Zhang, Y., Tang, T.T., Girit, C., Hao, Z., Martin, M.C., Zettl, A., Crommie, M.F., Shen, Y.R., and Wang, F. (2009) Direct observation of a widely tunable bandgap in bilayer graphene. *Nature*, **459**, 820.

382 Bostwick, A., Ohta, T., McChesney, J.L., Emtsev, K.V., Seyller, T., Horn, K., and Rotenberg, E. (2007) Symmetry breaking in few layer graphene films. *New J. Phys.*, **9**, 385.

383 Rotenberg, E., Bostwick, A., Ohta, T., McChesney, J.L., Seyller, T., and Horn, K. (2008) Origin of the energy bandgap in epitaxial graphene. *Nat. Mater.*, **7**, 258.

384 Sprinkle, M., Siegel, D., Hu, Y., Hicks, J., Tejeda, A., Taleb-Ibrahimi, A., Fèvre, P.L., Bertran, F., Vizzini, S., Enriquez, H., Chiang, S., Soukiassian, P., Berger, C., de Heer, W.A., Lanzara, A., and Conrad, E.H. (2009) First direct observation of a nearly ideal graphene band structure. *Phys. Rev. Lett.*, **103**, 226803.

385 Martin, J., Akerman, N., Ulbricht, G., Lohmann, T., Smet, J.H., and von Klitzing, K.A. Yacoby (2008) Observation of electron–hole puddles in graphene using a scanning single-electron transistor. *Nat. Phys.*, **4**, 144.

386 Rutter, G.M., Crain, J.N., Guisinger, N.P., Li, T., First, P.N., and Stroscio, J.A. (2007) Scattering and interference in epitaxial graphene. *Science*, **317**, 219.

387 Giovannetti, G., Khomyakov, P.A., Brocks, G., Karpan, V.M., van den Brink, J., and Kelly, P.J. (2008) Doping graphene with metal contacts. *Phys. Rev. Lett.*, **101**, 026803.

388 Zhou, S.Y., Siegel, D.A., Fedorov, A.V., and Lanzara, A. (2008) Metal to insulator transition in epitaxial graphene induced by molecular doping. *Phys. Rev. Lett.*, **101**, 086402.

389 Bekyarova, E., Itkis, M.E., Ramesh, P., Berger, C., Sprinkle, M., de Heer, W.A., and Haddon, R.C. (2009) Chemical modification of epitaxial graphene: spontaneous grafting of aryl groups. *J. Am. Chem. Soc.*, **131**, 1336.

390 Leenaerts, O., Partoens, B., and Peeters, F.M. (2008) Adsorption of H_2O, NH_3, CO, NO_2, and NO on graphene: a first-principles study. *Phys. Rev. B*, **77**, 125416.

391 Sessi, P., Guest, J.R., Bode, M., and Guisinger, N.P. (2009) Patterning graphene at the nanometer scale via hydrogen desorption. *Nano Lett.*, **9**, 4343.

392 Koshino, M. and McCann, E. (2009) Gate-induced interlayer asymmetry in ABA-stacked trilayer graphene. *Phys. Rev. B*, **79**, 125443.

393 McCann, E. and Fal'ko, V.I. (2006) Landau-level degeneracy and quantum hall effect in a graphite bilayer. *Phys. Rev. Lett.*, **96**, 086805.

394 Gierz, I., Riedl, C., Starke, U., Ast, C.R., and Kern, K. (2008) Atomic hole doping of graphene. *Nano Lett.*, **8**, 4603.

395 Guisinger, N.P., Rutter, G.M., Crain, J.N., First, P.N., and Stroscio, J.A. (2009) Exposure of epitaxial graphene on SiC(0001) to atomic hydrogen. *Nano Lett.*, **9**, 1462.

396 Schedin, F., Geim, A.K., Morozov, S.V., Hill, E.W., Blake, P., and Katsnelson, M.I., Novoselov, K.S. (2007) Detection of individual gas molecules adsorbed on graphene. *Nat. Mater.*, **6**, 652.

397 Wehling, T.O., Balatsky, A.V., Tsvelik, A.M., Katsnelson, M.I., and Lichtenstein, A.I. (2008) Midgap states in corrugated graphene: *ab initio* calculations and effective field theory. *Europhys. Lett.*, **84**, 17003.

398 Chen, Z., Lin, Y.-M., Rooks, M.J., and Avouris, P. (2007) Graphene nano-ribbon electronics. *Phys. E*, **40**, 228.

399 Brey, L. and Fertig, H.A. (2006) Electronic states of graphene nanoribbons studied with the Dirac equation. *Phys. Rev. B*, **73**, 235411.

400 Ezawa, M. (2006) Peculiar width dependence of the electronic properties of carbon nanoribbons. *Phys. Rev. B*, **73**, 045432.

401 Han, M.Y., Oezyilmaz, B., Zhang, Y., and Kim, P. (2007) Energy band-gap engineering of graphene nanoribbons. *Phys. Rev. Lett.*, **98**, 206805.

402 Datta, S. (1997) *Electronic Transport in Mesoscopic Systems*, Cambridge University Press, Cambridge, UK

403 Novoselov, K.S., Geim, A.K., Morozov, S.V., Jiang, D., Zhang, Y., Dubonos, S.V., Grigorieva, I.V., and Firsov, A.A. (2004) Electric field effect in atomically thin carbon films. *Science*, **306**, 666.

404 Bolotin, K.I., Sikes, K.J., Hone, J., Stormer, H.L., and Kim, P. (2008) Temperature-dependent transport in suspended graphene. *Phys. Rev. Lett.*, **101**, 096802.

405 Arnoult, W.J. and McLellan, R.B. (1972) The solubility of carbon in rhodium ruthenium, iridium and rhenium. *Scr. Metall.*, **6**, 1013.

406 Coraux, J., N'Diaye, A.T., Busse, C., and Michely, T. (2008) Structural coherency of graphene on Ir(111). *Nano Lett.*, **8**, 565.

407 Marchini, S., Günther, S., and Wintterlin, J. (2007) Scanning tunneling microscopy of graphene on Ru(0001). *Phys. Rev. B*, **76**, 075429.

408 Pletikosić, I., Kralj, M., Pervan, P., Brako, R., Coraux, J., N'Diaye, A.T., Busse, C., and Michely, T. (2009) Dirac cones and minigaps for graphene on Ir(111). *Phys. Rev. Lett.*, **102**, 056808.

409 Sutter, P., Sadowski, J.T., and Sutter, E. (2009) Graphene on Pt(111): growth and substrate interaction. *Phys. Rev. B*, **80**, 245411.

410 Varykhalov, A. and Rader, O. (2009) Graphene grown on Co(0001) films and islands: electronic structure and its precise magnetization dependence. *Phys. Rev. B*, **80**, 035437.

411 Wintterlin, J. and Bocquet, M.-L. (2009) Graphene on metal surfaces. *Surf. Sci.*, **603**, 1841.

412 Li, X., Cai, W., An, J., Kim, S., Nah, J., Yang, D., Piner, R., Velamakanni, A., Jung, I., Tutuc, E., Banerjee, S.K., Colombo, L., and Ruoff, R.S. (2009) Large-area synthesis of high-quality and uniform graphene films on copper foils. *Science*, **324**, 1312.

413 Kwon, S.-Y., Ciobanu, C.V., Petrova, V., Shenoy, V.B., Bareño, J., Gambin, V., Petrov, I., and Kodambaka, S. (2009) Growth of semiconducting graphene on palladium. *Nano Lett.*, **9**, 3985.

414 Shikin, A.M., Prudnikova, G.V., Adamchuk, V.K., Soe, W.-H., Rieder, K.-H., Molodtsov, S.L., and Laubschat, C. (2002) Synthesis of graphite monolayer stripes on a stepped Ni(771) surface. *Phys. Solid State*, **44**, 677.

415 Lahiri, J., Miller, T., Adamska, L., Oleynik, I.I., and Batzill, M. (2011) Graphene growth on Ni(111) by transformation of a surface carbide. *Nano Lett.*, **11**, 518.

416 van Bommel, A.J., Crombeen, J.E., and van Tooren, A. (1975) LEED and Auger electron observations of the SiC(0001) surface. *Surf. Sci.*, **48**, 463.

417 Berger, C., Song, Z., Li, T., Li, X., Ogbazghi, A.Y., Feng, R., Dai, Z., Marchenkov, A.N., Conrad, E.H., First, P.N., and de Heer, W.A. (2004) Ultrathin epitaxial graphite: 2D electron gas properties and a route toward graphene-based nanoelectronics. *J. Phys. Chem. B*, **108**, 19912.

418 Forbeau, I., Themlin, J.-M., and Debever, J.-M. (1998) Heteroepitaxial graphite on 6H-SiC(0001): interface formation through conduction-band electronic structure. *Phys. Rev. B*, **58**, 16396.

419 Qi, Y., Rhim, S.H., Sun, G.F., Weinert, M., and Li, L. (2010) Epitaxial graphene on SiC(0001): more than just honeycombs. *Phys. Rev. Lett.*, **105**, 085502.

420 Rutter, G.M., Guisinger, N.P., Crain, J.N., Jarvis, E.A.A., Stiles, M.D., Li, T., First, P.N., and Stroscio, J.A. (2007) Imaging the interface of epitaxial graphene with silicon carbide via scanning tunneling microscopy. *Phys. Rev. B*, **76**, 235416.

421 Emtsev, K.V., Speck, F., Seyller, T., and Ley, L., Riley, J.D. (2008) Interaction, growth and ordering of epitaxial graphene on SiC{0001} surfaces: a comparative photoelectron spectroscopy study. *Phys. Rev. B*, **77**, 155303.

422 Kitaura, R., Okimoto, H., Shinohara, H., Nakamura, T., and Osawa, H. (2007) Magnetism of the endohedral metallofullerenes M@C82 (M=Gd,Dy) and the corresponding nanoscale peapods: synchrotron soft X-ray magnetic circular dichroism and density-functional theory calculations. *Phys. Rev. B*, **76**, 172409.

423 Shinohara, H. (2000) Endohedral metallofullerenes, *Rep. Prog. Phys.*, **63**, 843.

424 Fuechsle, M., Mahapatra, S., Zwanenburg, F.A., Friesen, M., Eriksson, M.A., and Simmons, M.Y. (2010) Spectroscopy of few-electron single-crystal silicon quantum dots. *Nat. Nanotechnol.*, **5**, 502.

425 Kouwenhoven, L.P., Austing, D.G., and Tarucha, S. (2001) Few-electron quantum dots. *Rep. Prog. Phys.*, **64**, 701.

426 Kouwenhoven, L.P., Oosterkamp, T.H., Danoesastro, M.W.S., Eto, M., Austing, D.G., Honda, T., and Tarucha, S. (1997) Excitation spectra of circular, few-electron quantum dots. *Science*, **278**, 1788.

427 Ponomarenko, L.A., Schedin, F., Katsnelson, M.I., Yang, R., Hill, E.W., Novoselov, K.S., and Geim, A.K. (2008) Chaotic Dirac billiard in graphene quantum dots. *Science*, **320**, 356.

428 Kong, J., Franklin, N.R., Zhou, C., Chapline, M.G., Peng, S., Cho, K., and Dai, H. (2000) Nanotube molecular wires as chemical sensors. *Science*, **287**, 622.

429 Goldoni, A., Petaccia, L., Lizzit, S., and Larciprete, R. (2010) Sensing gases with carbon nanotubes: a review of the actual situation. *J. Phys.: Condens. Matter*, **22**, 013001.

430 Kauffman, D.R. and Star, A. (2008) Carbon nanotube gas and vapor sensors. *Angew. Chem.*, **47**, 6550.

431 Leenaerts, O., Partoens, B., and Peeters, F.M. (2008) Adsorption of H_2O, NH_3, CO, NO_2, and NO on graphene: a first-principles study. *Phys. Rev. B*, **77**, 125416.

432 Peng, S. and Cho, K. (2000) Chemical control of nanotube electronics. Nanotechnology (cursiv), **11**, 57.

433 Peng, S., Cho, K., Qi, P., and Dai, H. (2004) Ab initio study of CNT NO_2 gas sensor. *Chem. Phys. Lett.*, **387**, 271.

434 Valentini, L., Mercuri, F., Armentano, I., Cantalini, C., Picozzi, S., Lozzi, L., Santucci, S., Sgamellotti, A., and Kenny, J.M. (2004) Role of defects on the gas sensing properties of carbon nanotubes thin films: experiment and theory. *Chem. Phys. Lett.*, **387**, 356.

435 Zhang, J., Boyd, A., Tselev, A., Paranjape, M., and Barbara, P. (2006) Mechanism of NO_2 detection in carbon nanotube field effect transistor chemical sensors. *Appl. Phys. Lett.*, **88**, 123112.

436 Zhang, T., Mubeen, S., Myung, N.V., and Deshusses, M.A. (2008) Recent progress in carbon nanotube-based gas sensors. *Nanotechnology*, **19**, 332001.

437 Schedin, F., Geim, A.K., Morozov, S.V., Hill, E.W., Blake, P., Katsnelson, M.I., and Novoselov, K.S. (2007) Detection of individual gas molecules adsorbed on graphene. *Nat. Mater.*, **6**, 652.

438 Varghese, N., Mogera, U., Govindaraj, A., Das, A., Maiti, P.K., Sood, A.K., and Rao, C.N.R. (2009) Binding of DNA nucleobases and nucleosides with graphene. *Chem. Phys. Chem.*, **10**, 206.

439 Mohanty, N. and Berry, V. (2008) Graphene-based single-bacterium resolution biodevice and DNA transistor: interfacing graphene derivatives with nanoscale and microscale biocomponents. *Nano Lett.*, **8**, 4469.

440 Dreyer, D.R., Partk, S., Bielawski, C.W., and Ruoff, R.S. (2010) The chemistry of graphene oxide. *Chem. Soc. Rev.*, **39**, 228.

441 Kempaiah, R., Chung, A., and Maheshwari, V. (2011) Graphene as cellular interface: electromechanical coupling with cells. *ACS Nano*, **5**, 6025.

442 Cohen-Karni, T., Qing, Q., Li, Q., Fang, Y., and Lieber, C.M. (2010) Graphene and nanowire transistors for cellular interfaces and electrical recording. *Nano Lett.*, **10**, 1098.

443 Cai, H., Cao, X., Jiang, Y., He, P., and Fang, Y. (2003) Carbon nanotube enhanced electrochemical DNA biosensor for DNA hybridization detection. *Anal. Bioanal. Chem.*, **375**, 287.

444 Kamat, P.V. (2007) Meeting the clean energy demand: nanostructure architectures for solar energy conversion. *J. Phys. Chem. C*, **111**, 2834.

445 Nozik, A.J. (2002) Quantum dot solar cells. *Phys. E*, **14**, 115.

446 Green, M.A. (2002) Third generation photovoltaics: solar cells for 2020 and beyond. *Phys. E*, **14**, 65.

447 Green, M.A. (2006) *Third Generation Photovoltaics – Advanced Solar Energy Conversion*, Springer-Verlag, Berlin, Heidelberg, New York.

448 Nelson, J. (2003) *The Physics of Solar Cells*. Imperial College Press, London, UK.

449 Buonassisi, T., Istratov, A.A., Marcus, M.A., Lai, B., Cai, Z., Heald, S.M., and Weber, E.R. (2005) Engineering metal-impurity nanodefects for low-cost solar cells. *Nat. Mater.*, **4**, 676.

450 Tang, J. and Sargent, E.H. (2010), Infrared colloidal quantum dots for photovoltaics: fundamentals and recent progress. *Adv. Mat.*, **23**, 12.

451 Koleitat, G.I., Levina, L., Shukla, H., Myrskog, S.H., Hinds, S., Pattantyus-Abraham, A.G., and Sargent, E.H. (2008) Efficient, stable infrared photovoltaics based on solution-cast colloidal quantum dots. *ACS Nano*, **2**, 833.

452 Kongkanand, A., Tyrdy, K., Takechi, K., Kuno, M., and Kamat, P.V. (2008) Quantum dot solar cells. Tuning photoresponse through size and shape control of CdSe-TiO_2 architecture. *J. Am. Chem. Soc.*, **130**, 4007.

453 Kramer, I.J. and Sargent, E.H. (2011) Colloidal quantum dot photovoltaics: a path forward. *ACS Nano*, **5**, 8506.

454 Ma, W., Swisher, S.L., Ewers, T., Engel, J., Ferry, V.E., Atwater, H.A., and Alivisatos, A.P. (2011) Photovoltaic performance of ultrasmall PbSe quantum dots. *ACS Nano*, **5**, 8140.

455 Pattantyus-Abraham, A.G., Kramer, I.J., Barkhouse, A.R., Wang, X., Konstantatos, G., Debnath, R., Levina, L., Raabe, I., Nazeeruddin, M.K., Graetzel, M., and Sargent, E.H., Depleted-heterojunction colloidal quantum dot solar cells. *ACS Nano*, **4**, 3374.

456 Dai, Q., Wang, Y., Li, X., Zhang, Y., Pellegrino, D.J., Zhao, M., Zou, B., Seo, J.T., Wang, Y., and Yu, W.W. (2009) Size-dependent composition and molar extinction coefficient of PbSe semiconductor nanocrystals. *ACS Nano*, **3**, 1518.

457 Johnston, K.W., Pattantyus-Abraham, A.G., Clifford, A.P., Myrskog, S.H., and Hoogland, S. Shukla, H., Klem, E.J.D., Levina, L., Sargent, E.H. (2008) Efficient Schottky-quantum-dot photovoltaics: the role of depletion, drift, and diffusion. *Appl. Phys. Lett.*, **92**, 122111.

458 Huynh, W.U., Dittmer, J.J., and Alivisatos, A.P. (2002) Hybrid nanorod-polymer solar cells, *Science*, **295**, 2425.

459 Law, M., Greene, L.E., Johnson, J.C., Saykally, R., and Wang, P. (2005) Nanowire dye-sensitized solar cells, *Nat. Mater.*, **4**, 455.

460 Leschkies, K.S., Divakar, R., Basu, J., Enach-Pommer, E., Boercker, J.E., Carter, C.B., Kortshagen, U.R., Norris, D.J., and Aydil, E.S. (2007) Photosensitization of ZnO nanowires with CdSe quantum dots for photovoltaics devices. *Nano Lett.*, 7, 1793.

461 Robel, I., Subramanian, V., Kuno, M., and Kamat, P.V. (2006) Quantum dot solar cells. Harvesting light energy with CdSe nanocrystals molecularly linked to mesoscopic TiO_2 films. *J. Am. Chem. Soc.*, **128**, 2385.

462 Gruner, G. (2006) Carbon nanotube films for transparent and plastic electronics. *J. Mat. Chem.*, **16**, 3533.

463 Rowell, M.W., Topinka, M.A., McGehee, M.D., Prall, H.-J., and Dennler, G., Saricifti, N.S., Hu, L., Gruner, G. (2006) Organic solar cells with carbon nanotube network electrodes. *Appl. Phys. Lett.*, **88**, 233506.

464 Tyler, T.P., Brock, R.E., Karmel, H.J., Marks, T.J., and Hersam, M.C. (2011) Electronically monodisperse single-walled carbon nanotube thin films as transparent conducting anodes in organic photovoltaic devices. *Adv. Energy Mater.*, 1, 785.

465 Yang, S., Kalbac, M. Popov, A., and Dunsch, L. (2006) A facile route to the Non-IPR Fullerene $Sc_3N@C_{68}$: Synthesis, Spectroscopic Characterization, and Density Functional Theory Computations (IPR-Isolated Pentagon Rule). *Chem. Eur. J.* **12**, 7856.

466 Stephens, P.W., Mihaly, L., Lee, P.L., Whetten, R.L., Huang, S.-M., Kaner, R., Deiderich, F., Holczer, K. (1991) Structure of single-phase superconducting K_3C_{60} Nature 351, 632

Index

Inorganic Nanostructures: Properties and Characterization, First Edition. Reinke, P.

Published 2012 by WILEY-VCH Verlag GmbH & Co. KGaA